CRITICAL NORTHS

CRITICAL NORTHS

space nature theory

sarah jaquette ray
kevin maier
editors

UNIVERSITY OF ALASKA PRESS

Published by
University of Alaska Press
P.O. Box 756240
Fairbanks, AK 99775-6240

Cover design: University of Alaska Press, Fairbanks
Cover image: Broken pipe in Deadhorse, Alaska. From *The Last Road North* by Ben Huff.
Interior design: Rachel Fudge

Library of Congress Cataloging-in-Publication Data

Names: Ray, Sarah Jaquette, editor. | Maier, Kevin, editor.
Title: Critical norths : space, nature, theory / edited by Sarah Jaquette Ray and Kevin Maier.
Description: Fairbanks, AK : University of Alaska Press, 2017. | Includes bibliographical references and index.
Identifiers: LCCN 2016026148 (print) | LCCN 2016045179 (ebook) | ISBN 9781602233195 (paperback) | ISBN 9781602233201 ()
Subjects: LCSH: Arctic regions—Environmental conditions. | Climatic changes—Arctic regions. | Indigenous peoples—Arctic regions. | Human ecology—Arctic regions. | North (The concept) | BISAC: SOCIAL SCIENCE / Anthropology / General. | NATURE / General.
Classification: LCC GE160.A68 C75 2017 (print) | LCC GE160.A68 (ebook) | DDC 304.20911/3dc23
LC record available at https://lccn.loc.gov/2016026148

Contents

Acknowledgments

CRITICAL NORTHS WAS INSPIRED BY AN ASSOCIATION FOR THE STUDY OF Literature and Environment (ASLE) symposium “Environment, Culture, and Place in a Rapidly Changing North,” held at the University of Alaska Southeast in 2012. We would like to thank ASLE, especially managing director Amy McIntyre and then-president Joni Adamson; John Pugh, who was the chancellor at the University of Alaska Southeast; and Alison Krein, Virginia Berg, and Margaret Rea, our administrative support at UAS. They all enthusiastically supported the symposium in funding, time, website updates, program design, on-the-ground implementation, and genuine investment in this event and intellectual project.

The conversations held at that symposium precipitated many collaborations and projects, and so we would like to share the achievement of this book’s publication with all those who attended, spoke, and volunteered. We are indebted to our keynote presenters for shaping our discussions: Julie Cruikshank, Ernestine Hayes, Nancy Lord, and Ellen Frankenstein. In the spring semester leading up to the June symposium, we co-taught an upper-division undergraduate course and were inspired and awed by the students, who helped plan, orchestrate, and implement the symposium, in addition to contributing their own research as presenters. Their myriad commitments—from volunteering at the registration table to leading tours to catering the salmon bake to hosting guests—made hosting the symposium easier for us, but mostly it merged our research and teaching lives in meaningful ways.

The work of producing a book is mostly invisible to us scholars, so it is our pleasure to name the stalwart advocates of this project at the University of Alaska Press: James Engelhardt, Amy Simpson, Krista West, Rachel Fudge, and the Editorial Board. Eric Heyne deserves special recognition for his thorough reviews of the manuscript at multiple stages; thanks also to the blind peer reviewers for their thoughtful suggestions.

Thanks to photographer Ben Huff for the cover image. To us, the ironic hope suggested by the decrepit pipeline raises ambivalence about the petro infrastructure and gestures toward a central rhetorical goal of this volume. Huff evokes the irony in finding a landscape beautiful through the oil-shaped perspective of a windshield, or from the point of view of a road, a perspective we hope this volume elicits throughout. The line of the pipe also hints at the tensions between seeing the North as "wilderness" and seeing the North as resource. Finally, twenty-first-century pipelines inevitably highlight the transnational connections we raise in this book.

And thanks especially to the contributors who patiently responded to peer review, bent their disciplinary styles to fit this transdisciplinary project, and generally inspired us with their work.

—Sarah Jaquette Ray and Kevin Maier

Introduction

Approaching Critical Northern Issues Critically

sarah jaquette ray and kevin maier

THE "NORTH" HOLDS A POWERFUL AND OFTEN PARADOXICAL PLACE in the Western imaginary. As a region, it is often seen as simultaneously empty of life and full of promise, the home of indigenous communities yet utterly conquerable, pristine and desolate yet marked by a long human history, and isolated from the problems of civilization while at the same time the stage for a planet's climate drama. Symbolically, the North as resource has justified explorations both scientific and colonial, and continues to attract prospectors and adventurers in search of escape, transcendence, and fortune.

We see these northern notions in such popular texts and cultural productions as the Discovery Channel's suite of shows set in the North, including *Alaska: The Last Frontier* and *Deadliest Catch*, and the 2007 film *Into the Wild*. Literary icons like Jack London, John Muir, and Farley Mowat have immortalized the North as a space of (white) exploration, adventure, and natural wonder. Meanwhile, indigenous perspectives of the North remain marginalized in these narratives, despite figuring heavily as central to the trope of a "vanishing" and noble landscape. Northern literary and cultural productions have contributed immensely to colonial narratives of indigeneity and nation.

Recently, the North has taken on profound environmental significance. Images of melting ice caps and starving polar bears as well as internationally publicized debates about the Arctic National Wildlife Refuge (ANWR) place the North at the center of discussions about the urgency of the planet's environmental crisis. It is an "indicator region" for climate change. At stake are questions about developed nations' right to consume, economic progress, animal protection, indigenous sovereignty, and various tragedies of the commons. This transnationalization of the North is one story about the North that revises dominant perceptions of the region as a barren, isolated backwater. Even Sarah Palin understood this new view of the North when she claimed international diplomacy expertise by pointing to the fact that Russia can be seen from land in Alaska (a point that comedian Tina Fey famously rephrased to "I can see Russia from my house" in a *Saturday Night Live* skit). While this sentiment did not have the desired effect of increasing Palin's profile as a diplomat, it did exemplify a way of thinking about Alaska that those who dismiss her do not grasp.

This transnationalization of the North isn't necessarily a new story, of course, as battles over the North's resources have long been waged via distant and diverse political and economic networks. When we add climate change to the equation, however, the scale of our contemporary environmental problems is simply greater, the consequences higher for everyone—"here" as well as "from away" (from the perspective of the North). The developing world's desire for consumer electronics, rapid transportation, and constant high-speed connectivity have also upped the ante, as suggested by the slate of mineral extraction projects proposed, in development, and in production in the North. From the battle between corporations, nations, environmentalists, and First Nations communities over the tar sands in Alberta to the embattled Pebble Mine project proposed in the heart of key watersheds draining to Alaska's salmon-rich Bristol Bay to recent projections that the "best places to live"* in a climate-changed future will be in the North, the North's visibility in the past several decades has decidedly been due to its nature(s).

* See, for example, "Where Are the Best Places to Live If Global Warming Gets the Best of Us?" *Augusta Free Press*, May 15, 2015.

Less known but equally global in scale are some of the large mines in various stages of development in northern British Columbia, which pose similar global ecological problems. The political battles over the scores of mines in development and production in the region represent the transnationalization of northern environments even more directly. The Polley Mine is perhaps the most famous of the Canadian projects, as the tailings dam breach there made the social media rounds in 2014, and the story of inevitable downstream ecological destruction briefly made the news in the United States, but nearly a dozen Pebble-size projects are in the works. In most cases, these projects are driven by veiled corporate shells with unknown investors dispersed across the globe. Perhaps the more interesting transnational twist, however, is that while the United States Environmental Protection Agency will likely get the last say on Pebble, the United States has little regulatory authority on Canadian projects, despite the fact that many of the affected watersheds terminate in U.S. territory. In other words, these transboundary rivers along which the mines are being developed—the Unuk, Stikine, Taku, as well as others—provide critical habitat for the Southeast Alaska salmon that fuel a billion-dollar-per-year industry in the United States, but it is not clear to whom American fisherman might appeal to slow these huge mining projects that may impact their livelihood and the places they love.

While the fishermen and the environmental community face a long uphill battle against the many small corporations, perhaps the most interesting transnational development is the indigenous coalition lining up to push back. Once intimately connected, First Nations and Alaska Native communities are moving up and down long-traveled rivers, rekindling centuries-old relationships after a couple of generations of dormancy, in the process rerealizing the deep impact of the colonial border disputes that forced the Tahltan and Tlingit to work to sustain their ways of life with people in Ottawa and Washington, DC, rather than with neighbors with whom they share a long history and watersheds. Moreover, these First Nations and Alaska Native leaders are joining indigenous activists from the other side of the continent, learning from the similar experiences of direct action against the Keystone XL pipeline project and, more recently, the North Dakota Access Pipeline.

The Pebble and transboundary mine conflicts, the Keystone pipeline conflict, and the indigenous protest movement Idle No More all have transnational implications linking environments and people of the North with non-northern spaces and communities, both materially and symbolically. When we think of Bill McKibben's assertion that there is no "nature" anymore, which he espoused in his famous book *The End of Nature*, we think of the human impact of carbon on glaciers. We are reminded, moreover, that our desire for the latest iLife product requires mining on the scale proposed for these northern landscapes. In these debates, questions of identity and place, scale and ecosystem interconnectedness, nation and globalization all underscore long-held assumptions about the North as an anachronistic space set apart from most spheres of human activity. But if climate change is forcing the "American mind"[†] to acknowledge global interconnections between the North and everyplace else, then a critical examination of assumptions underpinning dominant conceptions of the North is needed in order to understand how tropes and narratives about the region have shaped and continue to influence environmental concerns.

What is at stake in defining the region? How have past definitions failed its environments and human and nonhuman inhabitants? As you will see in the essays herein, this volume seeks less to map out and describe a region (the North), as if that region's boundaries themselves were clear, than to consider what ideas about the North do to environments and peoples in and outside of the North, depending on how authors define it. In other words, what "counts" as the North is debatable, and a whole host of environmental issues—from identity politics to resource management strategies to species definitions—rest on how we define these boundaries. Because the North's landscapes, species, and peoples are at the center of debates about whether we are living in a new era, the Anthropocene, and because existing narratives of the North and the people who live there are often so inaccurate as to exacerbate social tensions and injustices, it is time

† We use this term to refer to Anthony Leiserowitz's compelling research on "Climate Change in the American Mind," which references Roderick Nash's book *Wilderness in the American Mind.*

for more ethical accounts of the North. The time for a new accounting is critical, and such a fresh accounting should be critical.

Hence the double meaning of the title of this volume—*Critical Norths.* While the plural *Norths* insists on the diversity of perspectives on what counts as northern, we also use the term *critical* here to signal urgency about these notions of the North, as the North increasingly becomes the canary in the coal mine of climate change. The problems that the North faces are critical, and while we want to be critical of how the North is deployed in alarmist discourse in our modern risk society, addressing environmental change by turning North may hold lessons and insights for communities in other regions. So we use the term *critical* for both its connotations: essays herein are critical about associations and definitions of the North, and they advocate a politics of urgency about the North.

This volume proceeds on the assumption that the North is not a static, discrete region, as classic geography suggests. Rather, we embrace the critical regional geographical theory espoused by scholars such as Doreen Massey, David Harvey, Krista Comer, Neil Campbell, and Susan Kollin that place is dynamic, layered, connected, and socially constructed. In this view, the North is as much a place as it is a set of cultural values and associations—what geographers call a *perceptual geography*. Following these critical regionalists, this volume espouses an idea of the "North" that dismantles regional geographical notions even as it operates within those territorial parameters. In these ways, we mean to be critical about the North, even as we accept it as a real and simultaneously reified region. That is, the essays included here critically engage, if not refute, accepted wisdoms about space, nature, and the North. For instance, they show that the North is a transnational space, not an isolated no-man's-land. They show that the North is not empty but rather historied and redolent of ongoing, resilient human and nonhuman life. They show that the North is symbolically heavy, even overdetermined, in the ways that it has authorized nation-building, patriarchy, humanism, and exploitation. They show that the North is defined by unique characteristics but that those characteristics are best understood as contingent. Ultimately, this collection considers the North as a productive lens—both material and discursive—through which to understand empire, national identity, gender, and the more-than-human world.

Organization of the Book

In some ways, the environmental discourse surrounding the North mirrors broader environmental rhetoric, as representations of northern environments tend to deploy the same tropes and tactics that we find in descriptions of other endangered environments. The first section of the book critically engages one of the most dominant—and arguably most successful—tropes in environmental rhetoric: the elegy. We have titled this section "The Vanishing North?" as a shorthand way of referring to (and questioning) the particularly northern strain of this literary commonplace. The vanishing North trope works like and is related to the vanishing Indian trope, and so this section argues for closer scrutiny of the material implications of the frame. While environmental literary scholars like Lawrence Buell and Greg Garrard have noted the efficacy of the elegy to imagine the end to stave off future destruction, most are also suspicious of the apocalyptic ramifications of this Chicken Little rhetorical tactic, and indigenous communities have long rejected the concept for its genocidal implications. Despite elegy's apocalyptic problems and colonialist investments, it remains a popular way to argue in favor of conservation of threatened northern environments and is the northern trope du jour, as evidenced in the popular notion of the North as the "last frontier." The authors in this section tackle this fraught narrative tool, ultimately rejecting it as not the most effective clarion to urgent, sustainable action. In this sense, this section launches the project with a measured approach to what it means to narrate "critical" urgency about the North in the first place.

The section opens with Elspeth Tulloch's meditation on three related texts that explore the decline of the Eskimo curlew. In her attentive close readings, Tulloch carefully details the problems in ecoelegy—a term she uses to mark a mode in which a future is imagined to invite prospective mourning. In her careful critique, Tulloch notes that we should be suspicious of elegy for a number of reasons, perhaps most notably how easily the mode adopts the rhetoric of the imperialist mode of writing about "vanishing" people who were not necessarily vanishing.

The second two essays in the opening section focus on Alaska rather than the Canadian Arctic, as Tulloch does. In his analysis of contemporary Alaskan literature, Will Elliott notes that climate change significantly complicates this wistful, even sentimental, approach to understanding the human relationship to nonhuman environments. As he puts it, acknowledging the reality of climate change means that the "pleasure of moral certainty, of definitive solutions, of a blameless way of being in the world, is off the table once we begin thinking beyond the human." For Elliott, today's environmental problems require rethinking the dominant environmental narrative forms, in particular moving beyond the traditional elegy. Elliott advocates "reaching out as widely as possible, transculturally, interdisciplinarily, to recompose Alaskan futurity." In Elliott's optimistic conclusion, he looks beyond the Western canon to suggest we might carefully borrow the representational power of Raven from Alaska Native traditions to find new modes and metaphors.

Allison Athens similarly advocates looking toward northern storytelling traditions to think about effective ways to calibrate environmental politics. Her essay opens with the story of a polar bear who wandered down to a subarctic Gwich'in Athabascan village, an anecdote she contextualizes with several other stories that situate the polar bear as a figure for climate change. Athens ultimately argues that "an attention to microstories and alternative storytelling paradigms suggest opportunities for continued relationships with polar bears and northern ecologies, even if, in a not-too-distant future, they exist in an altered form." Each of these three opening essays explores northern issues in detail to suggest both the limitations of the dominant ways we envision environmental issues as well as particularly northern alternative visions as we face even more uncertain times.

Although each of these authors scrutinizes the elegy through narratives about animals—which is not surprising, given the metonymical links between vanishing wild animals and vanishing wild spaces, as Sarah Whatmore has argued—we separate them from the next section, "Thinking with Northern Animals." While all six essays discuss animals, the second section focuses more on politics, law, and historical context than on affect and narrative. In the opening chapter of this section, Russell Fielding compares communities fighting to retain whaling rights through international treaties in Alaska and the Faroe Islands. Noting the problems with international law determining "aboriginal whaling rights," Fielding advocates for an approach to international whale hunting management that is conservation-oriented rather than identity-based. Fielding points out that in cross-cultural contexts indigeneity proves a slippery and inappropriate metric with which to manage whale hunting. Kurtis Boyer opens his chapter with an analysis of indigenous hunting as well, focusing specifically on Inuit hunting of polar bears. Like Fielding, he notes the limits of national and international legal discourse but develops his analysis to challenge conventional ways of conceiving of human-animal relationships more broadly. John Miller develops a similar critique of the normative notions that define appropriate relationships between humans and nonhumans, offering a compelling close reading of a quirky nineteenth-century exploration narrative. Miller's reading, as he concludes, "contests a normative imperialist history of arctic exploration and provides a more complex and self-reflexive view on the meanings of encounters between human and nonhuman creatures in the nineteenth-century Arctic." Consistent with critical moves in animal studies, Miller implicitly advocates for a version of posthumanism akin to the formulations in recent work by Donna Haraway, Cary Wolfe, and others, who challenge the boundaries between human and nonhuman. As the authors in this section and in the entire volume establish, this is a boundary that is especially fraught in northern contexts.

The third section, "Notions of North and Nation," also provides a strong critique of the colonial narratives about the North, focusing more directly on how these narratives impact human inhabitants of the region. Even more specifically, these essays highlight northern issues that challenge conventional definitions of the nation-state, emphasizing transboundary environmental, cultural, and imaginative rhizomes that are in part so fascinating because Alaska is divided from the contiguous United States by Canada. At times the lack of territorial contiguity creates more commonalities between Canada and Alaska than Alaska and the United States, complicating ideas of national belonging. Geographically and in terms of nation-building, then, the North challenges assumptions about territoriality and governmentality, space and power, identity and belonging. The essays in this section intervene in all these ways, though we have divided this section into two subsections, "Transnational Norths" and "Indigenous Norths." Although these sections are certainly related, as northern identities such as indigeneity are often the basis for thinking transnationally about northern concerns, the essays are distinct enough in their focuses that they warrant nuanced framing within the broader theme of "nation." Such a division allows thematic cohesion while not ignoring the myriad ways in which indigenous articulations about space and nature often directly interrogate the nation and nation-state. Whether they challenge existing national boundaries, reveal historical transnational connections, expose how the North is connected to—as opposed to isolated from—global networks of transportation, communication, and ecology, or scrutinize how some communities are excluded from the imagined community of the nation, all of these essays outline the ways in which notions of the nation underpin and are complicated in the North.

Each chapter in the first section on "Transnational Norths" points not only to the ways in which northern places are worked upon by global capital but also to the ways in which these landscapes and populations are implicated in larger global narratives of power, progress, and science. In the opening essay in this section, Chie Sakakibara reinforces this argument by linking communities in two very different parts of the North through their connection to whales. Sakakibara offers a powerful segue from the previous section on animals, especially Fielding's essay, as well as a bridge to the subsequent section on indigenous Norths. Similar to Fielding's approach to Faroese and Iñupiaq whaling communities, Sakakibara argues that the shared transnational cultural identity between the Iñupiat of Alaska and the Azorean in the North Atlantic fosters resilience in the face of the challenges posed by climate change. Providing a productive counterforce to the elegiac narrative scrutinized in the first section in her albeit-contested frame of "resilience," Sakakibara challenges conventional wisdom about essentialized indigenous identity and advances a nuanced theory about how transnational identifications might actually *enhance* place-based identity as a form of resilience in the face of the ongoing threats of colonialism and climate change.

Dovetailing with Sakakibara's theme of finding place-based identity through encounters with global forces, Janike Kampevold Larsen and Peter Hemmersam's essay argues for a place-based landscape art as a tool to resist transnational corporate interests. Critical close readings of community resistance efforts reveal how narratives of exploration and resource extraction by transnational corporations can overdetermine individual and community identity. Examining three communities along the Barents coast in Russia and Norway that are exposed to, as they put it, "extranational and multicorporate interests on a scale, and at a speed, that is unprecedented," the authors suggest that these towns exist "in a radical tension between their local, geologic, and microclimatic situations and global economic interests." To resolve this tension, the authors cite landscape art theory to advocate an approach to resolving political problems that is not just "site-specific, but also place-specific, sited both in space and time, and not only responding to projected scenarios but recognizing a multitude of local agencies, including that of local community and also that of landscape itself."

Carly Dokis similarly argues for local agency in environmental decision-making processes, analyzing the local response to environmental planning in the Sahtu Region of Canada's Northwest Territories. Dokis notes that governmental and corporate processes for determining the risk and impact of extractive projects too often consider only economics, not the local "spiritual, emotive, and ontological importance of the land and land-based practices."

In the final chapter in this section, Kyndra Turner frames two critical readings of literary imaginings of the North as a direct response to the rapidly emerging consensus that we are living in a new human-driven geological epoch. Offering a pair of models of critical reading in the Anthropocene, Turner suggests we should approach both Mary Shelley's classic novel *Frankenstein* and Richard Power's more contemporary novel *The Echo Maker* as interventions in the dominant narratives of exploration and power in an increasingly global imaginary brought on by climate change.

In the second subsection, "Indigenous Norths," we turn directly to northern notions of identity and representation, particularly indigenous identity and the complex politics of resource management in the North. In her contribution, Cheryl Fish reads two documentary films by Sami directors to argue that film provides a unique medium for exploring identity politics; they serve to counter the "imbalances that exist in the perception and representations of Sami by outsiders."

Susan Kollin also focuses on representation as a form of critique. She looks to the early-twentieth-century Hollywood career of Alaska Native actor

Ray Mala; in her close reading of both his life and his work, Kollin notes (echoing Sakakibara) that colonial narratives about cultural and place-based identities are often too narrowly conceived but that attentive readings can find unexpected instances of resistance in Ray Mala's work and engagement with Hollywood.

Connecting identity to narrative, the next chapter focuses on the power of oral history to contribute to our understanding of the human relationship to the nonhuman world. Daniel Monteith explores the confluences of oral history with a geologist's understanding of landscape change, arguing that scientists can benefit from studying traditional ecological knowledge. In addition, he demonstrates the power of interdisciplinary research methods and the pedagogical promises of working with undergraduates—particularly Alaska Natives—for complicating what has counted as knowledge about the natural world.

Finally, Margot Higgins applies a critical environmental historian's eye to the way Wrangell–St. Elias National Park and Preserve presents its mining and colonial past. She argues that these narratives erase contemporary indigenous influence in the area, an erasure that extends a tradition of omission from national park narratives and from the very landscapes themselves.

All of these last chapters present indigenous critiques of the North and of ideas of the nation that the North is deployed to support and, sometimes, subvert. We recognize that the organization of this volume privileges particular tropes and themes—including elegy, animal-human relations, and the nation-state—at the cost of equally valid alternatives, and we hope our readers will feel empowered to construct alternative groupings around other discursive formations (to use Foucault's term). Our sectional grouping does at least have the virtue of not privileging chronology, region, genre, or disciplinary approach, and thereby resists reinforcing some dominant Western paradigms. The fact that indigeneity and animals appear almost everywhere is certainly instructive, at least to us as editors. We hope that this volume as a whole offers a robust theorizing of interconnections between narrative, geography, human, and nonhuman in the North, and beyond. Most important, we hope that these essays reveal the complexities of the North—as an imagined region but also a material space with its own specific cultural and ecological histories and problems—that extend beyond the region.

Bibliography

Alaska: The Last Frontier. Discovery Channel. 2011–present. Television.

Buell, Lawrence. *The Future of Environmental Criticism: Environmental Crisis and Literary Imagination*. New York: Wiley Blackwell, 2005.

Deadliest Catch. Discovery Channel. 2005–present. Television.

Garrard, Greg. *Ecocriticism*. New York: Routledge, 2011.

Haraway, Donna. *When Species Meet*. Minneapolis: University of Minnesota Press, 2007.

Harvey, David. *Spaces of Hope*. Berkeley: University of California Press, 2000.

Into the Wild. Directed by Sean Penn. Paramount Vantage, 2007. DVD.

Leiserowitz, A., E. Maibach, C. Roser-Renouf, G. Feinberg, and S. Rosenthal. *Climate Change in the American Mind: October, 2015*. Yale University and George Mason University. New Haven: Yale Program on Climate Change Communication, 2015.

Massey, Doreen. *For Space*. New York: Sage, 2005.

McKibben, Bill. *The End of Nature*. New York: Random House, 2006.

Nash, Roderick. *Wilderness and the American Mind*. 4th ed. New Haven: Yale University Press, 2001.

Saturday Night Live. "Palin/Hillary Open." NBC. Aired Sept. 13, 2008. Television.

"What Are the Best Places to Live If Global Warming Gets the Best of Us?" *Augusta Free Press*. May 15, 2015. Blog.

Whatmore, Sarah. *Hybrid Geographies: Natures, Cultures, Spaces*. New York: Sage, 2002.

Wolfe, Cary. *What Is Posthumanism?* Minneapolis: University of Minnesota Press, 2009.

PART I

the vanishing north?

1

Whose Arctic? Who Cares?

Place, Responsibility, and Elegiac Purpose in the Eskimo Curlew Extinction Narrative

elspeth tulloch / UNIVERSITÉ LAVAL

IN "LOST DOGS, LAST BIRDS, AND LISTED SPECIES: CULTURES OF Extinction," Ursula K. Heise contextualizes scientific narratives about biodiversity loss within the tradition of stories about the deterioration of nature. She reminds us that typical narratives of decline focus on endpoints rather than new beginnings and simplify the way extinction can be understood. Relying mainly on tragic and elegiac modes, these tales either recycle a perennial unease with "modernization processes" or critique modernity's undermining of the relationship of humans with nature. Pointing to some of the limitations in conceptualizing species loss through these modes and drawing on Joseph W. Meeker's seminal work *The Comedy of Survival*, Heise explores the possibility of deploying an alternative approach to tell "'declensionist' narratives"—that is, at once comic and satiric. Following Meeker's lead, she argues that such an approach can propose a vision "not of the end of nature so much as its continually changing futures."[1]

When considering the fate of species in the North in an era of climate warming, this quest for an alternative narrative mode to tell such a story poses challenges.[2] Indeed, advocating a comic approach, which, as Meeker explains, presumes a paradigm of integration and adaptation, could be misconstrued as encouraging the minimizing or even ignoring of current and future losses of nonhuman species due to human activity. As such, its use could be seen as inviting the evasion of ethical engagement with these problems, Meeker's and Heise's arguments to the contrary. This possible sidestepping of issues is problematic, especially given that anxiety about irrevocable habitat change and the destruction of species unique to the North will no doubt persist. The reasons for this continued concern are multiple and well-known: the anticipated and accelerated alterations in northern environments, the concomitant and increased pressure for resource extraction, and the attendant human encroachment into once relatively sparsely human-inhabited areas. These realities are further complicated by the needs and claims of indigenous peoples with respect to resource use, access to traditional lifestyles, and related issues. Tensions over arctic sovereignty raise additional concerns about how to control or mitigate environmental damage. The relative rapidity with which some of this transformation is occurring understandably generates a deep sense of loss, especially among those who witness firsthand harm to ecosystems. For many wild species, adaptation will be complicated, difficult, and uncertain. Some, such as the ivory gull, face extinction.[3] As such, the situation invites the use of the conventional modes in question.

This chapter will not focus on the first of these modes—the tragic mode—given its partial connection with classical tragedy and the latter's problematic fostering of what Meeker qualifies as an arrogant conception of morality that sets humans up above their "natural environment and animal origins."[4] Rather, my argument considers the more contemplative elegiac mode, one that may reflect on, or invite the reader to reflect on, the loss the text describes.[5] Originally a term attached to certain poetry of lament and mourning, *elegy* has also been used to describe prose works with similar thematic concerns, such as those "dealing with a vanishing way of life."[6] When such texts are environmental essays, they can, in Timothy Morton's words, "fuse elegy and prophecy, becoming elegies for the future."[7] As such, "[e]cological elegy asks us to mourn for something that has not completely passed, that perhaps has not even passed yet."[8] The "anticipatory elegy" is a phenomenon Bonnie Costello also observes in ecopoetry, leading her to coin the term *ecoelegy*.[9] While pointing to current decline, the environmentally engaged elegy can thus evoke a sense of impending greater, indeed ultimate, loss in the future, which explains the mode's ubiquity in this age marked, as many biologists believe, by the sixth mass extinction. The evocation of extinction discourse in ecological elegies, through the lamenting of the future passing of entire species, has parallels with the extinction discourse that Patrick Brantlinger identifies in the "proleptic elegies" of imperialist texts, that is, elegies "mourning dying races before they had actually expired."[10] Aldo Leopold's essay "Marshland Elegy" in his classic *A Sand County Almanac* exemplifies the extinction-drenched ecological elegy. Describing the long-term degradation and near annihilation of a marsh, it concludes by envisioning the eventual demise of the migrating crane that once made this habitat its home. Leopold's elegy inspired at least one ornithological conservationist, George Archibald—originally from the Maritimes ("southern" Canada) and cofounder of the International Crane Foundation—to take up his life's work and eventually go North in quest of cranes.[11] His passionate response to an ornithological ecoelegy shows their power and invites their study, particularly as they deal with northern species and habitats far from the bulk of human populations affecting them, and especially species that migrate over southern space. In what follows, I attempt to answer the call for this type of work by examining iterations of a widely circulated yet little studied extinction narrative about a northern, or at least northern-attached, species—the Eskimo curlew. My aim is to elucidate how this narrative of a vanishing northern-attached species functions in popular culture, how it makes appeals to southern dwellers in North America, and how it acts proleptically by refocusing the elegiac mode from the past to future, as environmental elegiac literature often does. At a time when transregional, transnational, not to say global solutions are touted as key to addressing complex ecological problems, the desire to increase the sensitivity of populous and resource-consumptive southern dwellers to northern issues continues.

Extinction narratives drawing on unconventional modes will constitute welcome and necessary innovations in this process. Indeed, new, open-ended, and adaptive ways of conceptualizing the future of the North in this changing world, as Will Elliott and Allison Athens explore in their respective chapters in this volume, are vital for finding a path or paths in the uncertain way that lies ahead. Formulating ultimately life-affirming solutions will likely, however, require societies to move through a prolonged and messy coming-to-terms process. This will involve fluctuating between at least two states: on the one hand, dealing with loss, especially unnecessary, disruptive, or massive loss, and on the other hand, working restoratively for as richly diverse a future as possible while adapting to the future as it comes to be. Depending on how one defines it and how it is manifested, fighting against loss could figure in either part of this process. This complex grappling with events has parallels with the bereaved person's oscillation between "loss-oriented" and "restoration-oriented" efforts after the death of a loved one, a taxonomy psychologists Margaret Stroebe and Henk Schut proposed some fifteen years ago.[12] One part of this "working through" dyad does not preclude the other.

It remains germane then, during this period of decline and disappearance, to scrutinize how a frequently used mode makes appeals to the distant reader's or the distant spectator's sense of and responsibility for a particular faraway place and its common, threatened, and endangered species. In aestheticizing loss, the elegiac mode provides a cultural space for channeling and redirecting sadness, grief, mourning, and despair. As those seeking to foster ethically informed activism observe, sadness can serve to generate anger, which may translate into care, advocacy, and ecologically oriented action.[13] Indeed, the ecoelegy, whether in its poetic or prose form, functions in complex ways in this regard. As Costello, drawing on the work of R. Clifton Spargo explains,[14] "ecoelegy mourns a particular loss in . . . terms of anticipatory elegy, as if it could stop the loss and offer belated protection."[15]

Three interrelated elegiac renderings of the Eskimo curlew extinction narrative assign symbolic significance to this species, impart representations of the North and South, and make appeals to southern dwellers over the course of the second half of the twentieth century. It was during this period that the environmental movement and public interest in the North—its peoples and species—began to grow in contemporary culture. Catalysts for concern for endangered species included the founding of the International Union for the Conservation of Nature in 1948, which began compiling and releasing its Red List of Threatened Species in 1963. Almost a decade prior, in 1955, Fred Bodsworth (1918–2012), a Canadian journalist and naturalist, published his first novel, *Last of the Curlews*, an elegy heralding the impending extinction of the Eskimo curlew.[16] A bird reputed to have once turned skies black in its migrations, the Eskimo curley experienced a devastating and relatively sudden collapse in population prior to the turn of the twentieth century. Bodsworth's documentary novel was an expanded version of a novelette he had published to popular acclaim under the same title in *Maclean's* magazine on May 15, 1954. Selling more than three million copies, the novel helped regalvanize interest in conservation, something the decimation of the Eskimo curlew and the rapid decline of other birds had done in the United States and Canada at the turn of the twentieth century.[17] The novel has remained a staple among conservationist-minded readers ever since but has been largely ignored in literary scholarship.[18] In 1972, almost twenty years and several editions later (including a 1955 edition in the widely disseminated Reader's Digest Condensed Books series), the novel was adapted into an animated film for the ABC Afterschool Special by Hanna-Barbera Studios.[19] This broadcast coincided with the burgeoning environmental movement. In 1996, forty-one years after the novel's initial publication, the documentary film *The Barrens Quest* reappropriated the Eskimo curlew's story, showcasing Bodsworth himself.[20] The documentary exposes rekindling concerns about the effect of mining development on parts of the Canadian North.[21]

Extinction narratives often carry a dual signification: the real destruction of the actual species in question and the wider symbolic concept of the eradicative threat humans pose to nonhuman species in general.[22] The fate of the single male protagonist and his mate in Bodsworth's novel and its animated film adaptation and the fate of the Eskimo curlew as a species as recounted in all three texts carry more than these two connotations. As *The Barrens Quest* baldly concludes, "the story of endangered species invariably unfolds into the story of endangered spaces." In this light, the Eskimo curlew's extinction narrative can be read in the three texts as connoting both endangered northern species and northern spaces and, through the elegiac mode, as seeking to elicit the reader's desire to protect both.

However, when studied in chronological order of production, the three texts can also be seen as increasingly implicating northerners in the degradation of their environment, undermining the rhetorical appeal to southern dwellers to worry about the North's environmental future. This widening of the assignment of responsibility for species or environmental decline reduces the burden to be borne by those human populations most implicated in the decline. As such, it parallels to some extent Brantlinger's observations about the implicit stance of proleptic elegies that mourn vanishing races in imperialist rhetoric. He notes that these elegies found ways to blame the elegized peoples for their anticipated demise, even when the imperialist texts mourned the expected passing of those perceived to be doomed.[23] As I will show, this reworking of the burden of responsibility in the texts under study here also results, by the documentary, in a recontainment of place and a narrowing of ecoresponsibility. These aspects had been by comparison more widely assigned or shared in Bodsworth's novel and its animated adaptation. As an alternative strategy, the documentary attempts to rally the masses to the cause by appealing to ethnic identities dominant among southerners. Further, it widens the symbolic resonance of the Eskimo curlew. Together these tactics strongly direct the elegiac mode toward engagement with the future. This explicitly prospective stance brings the extinction narrative of one species squarely into the contemporary drama of current species in decline in the North. As such, the documentary draws on the full force of the anticipatory ecological elegy. This set of texts thus culminates with one that explicitly solicits outsiders to protect the region's threatened bio-uniqueness while it inadvertently disengages them. One is left wondering, then, about the efficacy of the mode, given the history and geopolitics of the region. A more detailed review of each text will demonstrate my arguments.

Last of the Curlews (1955)

Like many extinction narratives, Bodsworth's novel *Last of the Curlews* is constructed around a flagship species, in this case one that was once abundant, numbering by some accounts in the millions,[24] and was capable of especially grueling feats of migration. To heighten the reader's concern for the species, the narrative focuses on one of its last remaining males—if not the last[25]—as he attempts to find a mate. Over the course of a yearlong cycle, he migrates from the North to the southernmost tip of South America and back. In the warmer climes of the South, he finally locates the elusive female—perhaps the last remaining one—and heads North in the hope of mating with her. A farmer shoots her before they make it to the arctic breeding grounds. She succumbs to her wounds, the male protectively at her side, the mating cycle never completed. Driven by instinct, he returns to the Arctic to stake out his territory and await a new mate.

The chronologically told migratory narrative of this largely solitary male, accompanied only partway by a female, is intercut with excerpts from historical texts, each titled "The Gauntlet." These succinctly but effectively document the massive slaughter to which the Eskimo curlew as a species was subjected in its past migrations, particularly in late-nineteenth-century North America. The tightly constructed novel thus relentlessly insists that this single male curlew is solitary mainly because humans engaged in the deliberate destruction of the species over a brief span of decades. Along much of the curlews' migratory route, the descendants of settlers wantonly hunted them down and, when the opportunity presented itself or their shot was depleted, clubbed exhausted migrants to death. The demise of the female at the hands of the farmer hints at two other major reasons for the collapse of the species' population, one not directly addressed in the novel: that is, the supplanting of native prairie, the source of the species' natural food supply during its northerly migration, by cropland, and the annihilation of the Rocky Mountain grasshopper.[26] The pair's need for sustenance in what was once native prairie brings them too close to humans; the birds follow plows churning up the grubs they need to carry on, inciting, for nebulous reasons, the farmer's lethal attack.[27]

The narrative vaguely sets the migratory starting point for "thirty-odd shorebirds," like the Eskimo curlew, somewhere in the "Canadian Arctic."[28] It attaches the Canadian national adjective to the Arctic only once, however. For the other many times the Arctic per se is mentioned in the novel, it is never specifically associated with Canada. The Arctic remains largely designated as an a-national space, denominated by terms such as "the tundra" and "the barrens." The exact site of the desired nesting ground of the curlew protagonist is never given.[29] This is a striking omission, given that the curlew's southernmost feeding grounds are specifically situated in the Argentine portion of Patagonia and that other toponyms are regularly mentioned throughout the narrative to situate the reader along the curlew's migratory route.

Thus, while two narrative clues suggest that the curlew likely starts its migratory loop in the Canadian Arctic and seeks to return to territory located in Canada after passing the winter in the south,[30] I do not share Lee Frew's reading that these barest of clues serve "to appeal to a specific settler nationalism,"[31] at least not strongly. Rather, I read them as factual acknowledgment of where the species was actually known to nest.[32] What stands out more prominently in the text is the juxtaposition of the multiple generic references to the Arctic (noun) with the string of references to specific Canadian and American settler spaces to the south.[33] Together these propose a lethal North-South divide. The Arctic is idealistically presented as a region devoid of wantonly destructive humans,[34] with the settler spaces to the south as sites historically filled with sport and market hunters and blanketed with agricultural landscapes peopled with enemies of the avian world.

This particular species of curlew evokes the North as a nonnational region inhabited by First Peoples. It carries this complex set of connotations by virtue of its dated common name (the Eskimo curlew), the occasional translation of its Latin nomenclature *Numenius borealis* as "spirit of the North,"[35] and its general historical range (mainly the Northwest Territories in Canada but also possibly Alaska).[36] Simultaneously, the Eskimo curlew represents transnational and transcontinental movement, one embracing the northern and southern hemispheres. Its migratory direction is either toward the North (in the spring) or the South (in the fall) and is repeatedly mentioned as such in the narrative. Indeed, its range is specifically described as stretching from "the very northernmost to the southernmost reaches of the mainland of the Americas."[37] All the destructive forces that threatened the bird lie between, with those in the United States and Canada described using brutally depressing statistics. Both settler nations are thus indicted in the species' fate. While historically much of the hunting of the Eskimo curlew during the spring migration occurred on the American plains, the novel specifically indicts Canada in the species' final destruction, setting the killing of the female on the mechanized "Canadian prairie."[38] Having the female die so close to its nesting destination in the Arctic heightens the irony but also reminds southern-dwelling Canadian readers, arguably Bodsworth's original audience, of their current connection to migrating northern species as well as earlier Canadians' part in this particular bird's decimation.

Although a migratory creature, the Eskimo curlew's home base is firmly presented as the North. In the Arctic lie its nesting grounds—the land in which it would raise its young if it had any. All along the curlew protagonist's flight south, comparisons to the Arctic are made, and the pull of the Arctic eventually calls it "home."[39] The Arctic remains the ultimate reference point, wherever the bird finds itself. It is therefore a multiterritory creature, linked inexorably to the North, dependent not only on its own survival instincts and skills but also on the goodwill of southern denizens to let it pass unmolested.

One may thus interpret the Eskimo curlew as a metonymy for the North and the documents about the species' slaughter as a cautionary tale alluding to the possible fate of other aspects of the North's ecosystem intimately attached to or dependent on the South. Read this way, the novel suggests that the destruction of parts of the region's environment or nonhuman life will be well under way before those outside the region fully understand what is happening. By the time they do, it will be too late. As such, the novel functions as an ecoelegy, with certain sections heightening this anticipatory perspective by drawing on the building evidence from the past to suggest the future must be yet bleaker. For example, the novel includes an excerpt from the Proceedings of the Boston Society of Natural History (1906–7), which, in noting the sudden collapse of the Eskimo curlew's population, concludes that the species is "on the way to extinction."[40] All subsequent excerpts from ornithologists' and other scientific reports offer only further variations on the fear of the bird's extinction. One or two hold out hope for the regeneration of the species, based on a single sighting of either a pair or a solitary bird. But the story of the solitary, *fictional* curlew protagonist dominates the novel. The focus on *his* story suggests the species can only live in the imagination, in a migratory loop of ever-diminishing hope.

While after the female's death the last imagined male carries on, impelled by the instinctive desire to find another mate, we grasp the situation's dramatic irony. Versed by the end of the novel in the numerous reports about the wanton killing of the species, we understand the bird's desire to mate will likely remain unfulfilled, and be nonregenerative for the species even if it is. As much as we may root for the consummation of the curlew protagonist's desire, and as much as we may fiercely hope, against all realistic evidence, for the species' survival, we ultimately must despair. Indeed, by initially drawing on the rhetoric of the proleptic elegies of imperial discourse, the narrative falls, perhaps inadvertently, into a convention of hopelessness. In so doing, it makes the problematic concept of the "vanishing race" a trope for the dying species. This troping occurs even though the narrative remains firmly trained on the curlew story from start to finish and the rhetoric ornithologically factual.

Specifically, at its outset the narrative sets us up to mourn the passing of a "dying race,"[41] with the male curlew as one of its last representatives. To bring the species' death knell closer as the narrative draws to its conclusion, the female is killed just before her eggs can be fertilized. This penultimate scene, with its ambiguously anthropomorphic description of frenzied carnality, carries, in Lee Frew's reading, the hint of victim blaming and sexualized stereotyping found in imperial discourse about so-called doomed races.[42] The hormonally driven wild creature is extinguished in a moment of physical abandonment. The imperial/colonial trope of the dying race and its connotations of racial superiority are arguably transferred to human–nonhuman relations, in which "superior" human firepower outdoes the bird in the end.

That said, the burden of responsibility for the species' demise remains, as the narrative is constructed, on humans, inviting reflection on environmental ethics. At several points in the narrative, characteristics that made the species vulnerable to human attack are noted. Although this factual acknowledgment could be interpreted as introducing the "blame the victim" strand of early extinction discourse about species,[43] humans are the ones indicted. They are repeatedly shown to be taking advantage of the species' vulnerabilities for reasons unrelated to human survival. Moreover, by setting the protagonist curlew's journey in recent times, the narrative emphasizes the ecological and nonhuman rather than the imperialist and human strand of elegies that emerged with the advent of modernity in North America. As the reader learns at the outset, all living memory among contemporary, local, indigenous people of the great returning flocks is already dead. By thus evoking the generations-long period of time since the migrating bird darkened the skies, the narrative intimates the immense odds against it ever recovering its former numbers, should it—improbably—even survive as a species. Ultimately, given the narrative's nonhuman focus, the novel can be read as bearing a contemporary ecological warning for the region as a whole.

Indeed, read as symbolizing the fate of the North's environment, the narrative trajectory of the curlew protagonist suggests that the North will not only be unable to regenerate past losses but will be left alone to face its fate, like the curlew after its mate's death. A tale of ecotragedy and ecodespair, the text's permeation with biorealism ultimately dims, if not forecloses on, any hope for renewal and new beginnings. The novel thus operates mainly as an ecoelegiac accounting of the final phase of (a future) loss. However, the framing of the ecoelegy within imperial/colonial elegiac tropes undercuts impetus to action, inadvertently letting readers to the south (whom some historians have conceptualized as the North's colonizers[44]) disengage from the curlew's fate. The incorporation of historical excerpts on the gauntlet of slaughter in the narrative serve to denounce the effects of settler modernity. Nevertheless, the underlay of the dying-race paradigm, steeped, as Brantlinger observes, in self-justifying imperialist and expansionist ideology,[45] allows the curlew's demise to be understood and ultimately excused as a sad yet inevitable passing—the price of progress. Even as modernity—symbolized by the tractor-driving farmer-hunter—is shown to be the ultimate culprit, the critique is tainted by the narrative's initial allusion to the mode as implemented in imperial and settler contexts. This allusion carries connotations of resignation and acceptance.

The Last of the Curlews (1972)

The Last of the Curlews, the animated film adaptation of Bodsworth's novel, was produced by Hanna-Barbera Productions for the first of the ABC Afternoon Specials, airing on October 4, 1972. It garnered several awards, including the Emmy for Outstanding Achievement in Children's Programming in 1973, and was cherished by its producers, William Hanna and Joseph Barbera, and its production designer, Iwao Takamoto, as a high point in their respective careers.[46] Part of the animated film's appeal lies in its attention to natural background imagery. The nuanced care given to rendering color and detail is striking when compared to the minimalist aesthetic characterizing television cartoon series favorites like *The Flintstones,* for which Hanna-Barbera Productions were by then famous. In the television cartoon industry of the period, the depiction of the physical environment of animated characters was largely sacrificed and bodies simplified to save time and cost. The background imagery of this particular Hanna-Barbera production, however, harkens back to the realism of Disney's feature-length animations, an aesthetic Takamoto had learned to master when he had previously worked on Disney productions. The artistic attention given to the Eskimo curlew's natural world thus bespeaks the keen interest of both the producers bankrolling the production and the production designer in celebrating the narrative's natural context. As Barbera affirms, "[T]he cartoon . . . is . . . full of images of flight, and of nature in its loveliness and its violence. The special was a labor of love."[47]

The environmentally rich background imagery serves as an enticing, albeit silent, counterpoint to the film's heavy-handed environmental didacticism. Hammered home by the film's voice-over narrator, Lee Vines, and the maudlin and stilted introduction of a boy learning about the harsh realities of hunting, this morally overbearing tone contrasts with the neutral tone of the novel.[48] Nonetheless, this otherwise strident "enviro-toon"[49]—or cartoon with an environmental message—evocatively uses its imagery to present its child audience with the varied transnational environment of the bird while subtly inviting the spectator to develop an appreciative eye for the natural world. This tactic heightens reasons for the spectator to care about this world, its species, and their impending loss. Like the illustrations in the original novel (drawn by accomplished bird illustrator T. M. Shortt in the Canadian editions and partially adapted by naturalist illustrator Abigail Rorer in the American Counterpoint edition), the film's background imagery breaks place open, visually reveling in environmental diversity, multiterritoriality, and, implicitly, transnationality.

However, in further individualizing the hunting culture, which is blamed in the novel and its animated film adaptation for decimating the species, the film singles out ordinary northern dwellers as among the, if not *the*, primary and current stewards of migrating arctic birds. The nineteenth-century mass sport and group market hunting, fueled by capitalist forces, is indicted in the novel, with the farmer-shooter introduced at the end to represent the perpetuation of a venatic culture in the agrarian ethos. The film, however, places considerable emphasis on a northern-based, one-man/one-animal hunting culture while attempting to discredit the underlying self-justifying assumptions the film associates with that culture. To build its stance, the adaptation intersperses throughout the film three one-rifle-to-one-animal hunting scenes involving a father and son (characters nonexistent in the novel) while incorporating a single historical sequence about nineteenth- and twentieth-century sport hunters early in the film. This sequence replaces the gauntlet of accounts of this activity that are intercut throughout the novel's main story line of the male curlew's quest for a mate. The accumulation of these accounts in the novel constructs a stereotypical divide between a slaughterous, wasteful, and greedy human South and a pristine, fragile, and wild North (which, except for hunters in Labrador, does not act to eradicate the Eskimo curlew). By contrast, the film specifically draws contemporary hunters of the North into the paradigm. In so doing it implicitly, if unintentionally, lets southern dwellers off the hook for contemporary conservation.[50]

The boy is shown learning about hunting from his father in a northerly although jurisdictionally unidentified environment. In their second appearance, the two are wearing traditional snowshoes and tracking a buck through deep snow in an evergreen forest. The film suggests that northern hunters—in particular—are the ones who must now learn to respect conservation laws. The son is portrayed gradually having to disassociate himself from some of the teachings of his father, who voices an erroneous understanding of the actual size of some migrating bird populations, among other misconceptions. The father preaches that human hunters, like nonhuman predators, help ensure the balance of nature to justify, it seems, his subsequent questioning of laws against shooting migratory birds. These laws, he implies, have allowed an explosion in the bird population. The voice-over narrator intervenes with a corrective, explaining that currently "there's only one" Eskimo curlew down in Argentina, desperately seeking a mate. The film's title, which adds the article "the" to the beginning of the source text's original title, emphasizes that this male along with the female mate it will temporarily find are indeed "the last" Eskimo curlews, removing the ambiguity implied by the novel's title. The film starkly concludes: "Once there were many thousands of Eskimo curlews; then there were two; now there's one. Soon, there will be none."[51]

The film carefully keeps the endangered natural world nation-free. It does not suggest whether the farmer who kills the female resides in the United States or Canada, merely referring to the "rich farmlands of the north." In this way the film situates the venatic behavior of contemporary farmers and hunters regionally and avoids imparting national blame or specifically accepting, as in the novel's version of this scene, a nation's part in the final downfall of the species. Similarly, the animation retains no specific reference to the "Canadian Arctic," focusing instead on the novel's more commonly generalized reference to the "Arctic." The relationship of a shot of the migrating bird overlaid over a crude map of the northern and southern hemispheres is too imprecise to suggest a national territorial starting point for the curlew. That said, the entire northern hemisphere, including Alaska, is on the map. The film thus invites spectators in both the United States and Canada to associate the Arctic with their respective northern jurisdictions, permitting a dual patriotic appeal. Cartographically allowing that the species may emanate from an American jurisdiction subtly encourages the largely American child audience to care about the bird.

This type of dual national territorial designation for the Eskimo curlew's nesting home is found in some publications with conservation aims for the species. For example, *Endangered and Threatened Animals of Texas: Their Life History and Management*, published in Texas, assuredly claims that the Eskimo curlew has its "nesting grounds in the Alaskan and Canadian Arctic"; most scientific reports, including ones by Alaska wildlife officials, are, however, somewhat more circumspect, listing Alaska only as a possible but unconfirmed nesting territory.[52] It is relevant to note here that Texas is the site of the last confirmed live-bird sighting of the Eskimo curlew (Galveston Island in 1962) and of several unconfirmed sightings before and since. It is thus to the advantage of conservation efforts in that state to designate the Eskimo curlew as an animal "of Texas," as the book's title suggests, and to affirm, as much as possible, that it nests in Alaska in order to reinforce patriotic reasons among Texans for protecting it. From the American conservation perspective, it is helpful for southern dwellers as far as the Deep South to feel that the species is in part "theirs" if they are to participate in its protection. The Hanna-Barbera production subtly plays this card, changing the novel's reference to "the swampy shores of Louisiana and Texas" to "the welcoming shores of Louisiana and Texas" in the narrator's description of the curlew pair's migration north. In sum, while didactically attempting to teach northern hunters to become more conservation conscious, the film gently invites southerners to embrace the endangered species when it crosses southern territory. A largely elegiac tale of past loss is thus embedded with contemporary conservation messages that transcend national space, overtly commenting on hunters in northern climes and more subtly appealing to North American dwellers farther south.

The Barrens Quest (1997)

Building on the interest that Bodsworth had generated in the Eskimo curlew's fate—and incorporating Bodsworth as an expert witness—the forty-four-minute documentary film *The Barrens Quest*, directed by Peter Blow, explores concerns about the anticipated environmental impact of diamond mining in the Northwest Territories in Canada in the 1990s. As such, the film brings the retrospective elegiac mode into contemporary regional development debates. Bodsworth's novel and its animated film adaptation focus on the human contribution to the likely extinction of this single flagship species in a time well past. The documentary film uses instead the brutal role that nineteenth-century market hunters played in the decimation of the Eskimo curlew as a species to sound the alarm about contemporary human disruption and destruction of the fragile ecosystem that the likely defunct species once called home. The elegiac mode is firmly redirected to the future and attempts to serve as a catalyst to action. In the half decade leading up to the shooting of the film (1991–1996), a large portion of the tundra had become the site of aggressive diamond mining exploration, which, in the words of the film's narrator, was "one of the largest mineral rushes the world has ever seen." The film cautions that the resulting increased human encroachment into wild habitat—especially the infrastructure supporting the mining activity coupled with hunters' easier access to herds—would threaten a delicate ecosystem. One biologist, Anne Gunn, warns that large numbers of wild creatures, like those in the Bathurst caribou herd, are no protection against population collapse, pointing to the case of the once-abundant Eskimo curlew.

Blow's documentary appeals specifically to Canadians to protect their threatened northern wilderness, as symbolized by flagship species like the Eskimo curlew. This strategy contrasts with the Hanna-Barbera production, which gestured to a dual patriotic appeal in arctic bird conservation by suggestively widening the Eskimo curlew's known Canadian arctic nesting territory to encompass the Alaskan Arctic. Blow's appeal to Canadian patriotism is unsurprising given that the film was originally destined for broadcast on Canadian national television by the Canadian Broadcasting Corporation, with further distribution by the National Film Board of Canada. Cindy Kenny-Gilday, a Dene environmentalist from the Northwest Territories who appears near the beginning of the film, asserts, "Canadians have to take the responsibility for what we have here in the tundra; it's every Canadian's heritage; it's every Canadian's tundra: one of the last wildernesses in the world, one of the last roaming herd of caribou in the world." Her call speaks to the meshing of cultural identity with emblematic species—with the loss of the latter weakening the former, a fear of which is common to extinction elegies, as Heise observes.[53]

Using the clip of Kenny-Gilday calling this wilderness "tundra" and the caribou it supports a "gem," the filmmaker builds a complex trope, attempting to implicate Canadians culturally and politically in the fate of the North. The film uses diamonds to symbolize both the capricious desires of a consumer culture and the beauty and preciousness of the fauna and flora of a profuse yet vulnerable ecosystem. The quest for the first type of gem is shown to threaten the second. At the same time, the wild gems of today recall a lost wild gem of yesterday, which the film visually constructs as the Eskimo curlew. Its fate, according to the narrator, "symbolizes the dangerous future confronting all the creatures of the tundra," an idea Bodsworth's novel only implies.

The film blunts the edge of its conservationist lesson to Canadians, however, by pulling in two directions. On the one hand, Kenny-Gilday's appeal to Canadians invites them to assume responsibility for their jurisdictional wilderness. On the other, the documentary's singling out of American hunting activity in the destruction of the Eskimo curlew lets Canadians and their pre-Confederation ancestors off the hook for any part they might have had in the bird's demise. The film recalls big eastern cities'—notably Boston's—craving for the Eskimo curlew as "a highly prized delicacy" while ignoring hunting sprees in Labrador, something Bodsworth's novel does record. Reference to massive hunts "out west," qualified by "the final carnage . . . on the American plains in the spring," keeps the focus trained south of the border on the slaughter perpetrated by the relatively numerous and aggressive market and sports hunters there. This strategy evades the issue of habitat destruction, which, already widespread in the United States, would have been Western Canada's most significant, although belated and relatively lesser contribution, to the bird's final undoing.[54] Similarly, the film's focus on the bid of the huge Australian company BHP (Broken Hill Proprietary Company) for permission to mine the "pristine tundra" for diamonds highlights the impending intrusion into the wilderness as foreign. This emphasis on the alien in the destruction of wild species and habitat can be construed as a backhanded form of Canadian nationalism.

The tendency to indict nonnational forces muddies the message to southern-dwelling Canadians, especially when coupled with the film's understandable focus on the debates surrounding the region's economic developmental desires and the Aboriginal people's conflicted position (as traditional hunters of caribou and seekers of more sovereignty, jobs, and prosperity). Effectively, southern-dwelling, especially non-Aboriginal, Canadians are allowed to divorce themselves from their own economic implication in disruptive resource extraction and to leave the struggle over the future of the arctic wilderness to local forces already so engaged. In Bodsworth's novel of a migrating bird, place had been opened up, a strategy built on its animated film adaptation. In the documentary film, place is returned to a single, more contained region, governed by the familiar nationalist paradigm of Canada versus the outside.

This interest in a single place is complicated by a focus on the region's internal struggle over whether and how to proceed with economic development, which implicates the North in at least the partial destruction of its own wild territory. The seductive claims of modernity are shown to divide northern peoples, leaving Aboriginal and non-Aboriginal people alike in three camps: those who reject the sacrifice of wild nature and traditional lifestyles; those who advocate development; and those who seek a negotiated space for wildlife and traditional ways while allowing modernization via mining. The film's rhetorical construction allies itself with the first. Kenny-Gilday's appeal to the patriotic desire of Canadians to protect their natural heritage is particularly articulate and heartfelt. In spite of it, the film's narrative construction arguably leaves southern-dwelling Canadians, already physically far from realities they will likely never see, in the largely unchallenged role of disassociated spectators. This is especially likely in the case of disengaged non-Aboriginal people.

Perhaps the wider, albeit conventional, appeal in the film is that made to the conservation- or preservation-minded spectator's sense of pathos and ethos through the film's reworking of the quest narrative in Bodsworth's novel. Rather than building its suspense around an Eskimo curlew's ultimately fruitless quest for a mate, the documentary interweaves several quests for several gems, real and figurative. A lone biologist's ultimately unsuccessful ten-year quest for the elusive Eskimo curlew—one of the North's lost wild gems[55]—threads through the competing quests of the mining companies, the region's governments and peoples, and the wildlife conservation community for their respective contemporary "gems." Drawing on a plot device common to other extinction narratives, the film follows the efforts of biologist Joachim Obst as he traces the known past nesting sites of the bird. All are empty of any evidence of the curlew's continued existence. As the film's title bleakly puns, the region's economically desperate quest to cash in its resources in exchange for employment and economic growth may ultimately leave it dry. The barrens, suggests the documentary, may end up barren—a sterile place—bled of its gems, both the diamonds and the wild creatures that flourished above them. Cautionary and elegiac, the film hints at the apocalyptic.

If the film's narrow regional focus may inadvertently reinforce southern apathy, the documentary film's use of noninsider witnesses reaches out to a non-northern audience. Their appearance invites a sense of vicarious belonging to the North or at least vague identification with northern proponents by swaths of southern-dwelling Euro-Canadian spectators. Several of the scientists who are either interviewed or filmed testifying before one of the environmental hearings are not originally locals, something evident from their accents: Obst emigrated from Germany, Gunn from Britain; another, David Schindler, is a Canadian citizen of American origin working at the University of Alberta; a fourth unnamed witness passes as a southern white Canadian who fell in love with the tundra after countless hours walking it. All speak passionately about their concerns for northern wildlife and habitat. Obst, who also interviews Aboriginal people to determine their present and past knowledge of the Eskimo curlew, goes so far as to take sides in the environmental-development debate, publicly censuring his former employer, BHP, at a hearing for insufficient respect for Aboriginal people and northern species. His efforts as a mediating figure may help white, southern-dwelling spectators feel closer to those Aboriginal people who, elsewhere in the film, raise their voices against the destructive aspects of development.

The film thus buttresses the pan-Canadian call of concerned First Peoples to protect the wild species and their tundra habitat (made by Kenny-Gilday in her opening adjuration and by other First People later in the film) with Euro-Canadian examples of support for this cause. Indeed, the film attempts to forge the notion of an inclusive bioregion—one that reaches out to non-northerners. It does so by making appeals to pan-Canadian nationalism then showing the area to be populated by a mixture of peoples advocating for the welfare of unique wild northern species and drawing on dominant ethnic markers of non-Native Canadians to the south to encourage a populous potential political ally outside the region to identify with the region.

This focus on a current state of openness sidesteps, however, the complex colonial and neocolonial history of the region, one that led to the intensive exploitation of its resources, the influx of non-Aboriginal peoples, and the attendant acceleration of social and habitat disruption.[56] That said, the film's rhetorical construction assumes a critical position. It tacitly suggests that the undermining or loss of the First Peoples' traditional hunting-based lifestyle, together with the anticipated loss of the wild species, notably the caribou that supports it, will have at least two consequences. It will change, perhaps irrevocably, the lives of those immediately affected, and it will simplify the multicultural fabric of the nation. Embedded in the elegy of the northern species and their lands is the proleptic elegy of both a people's way of life and a portion of a nation's identity.

One could argue that the documentary filmmaker and the production apparatus supporting him are indulging in an intranational form of what Renato Rosaldo termed "imperialist nostalgia."[57] Based as they are in southern Canada, they are arguably complicit in the contested disruption of the tundra, its peoples, and its life-forms. They could be seen as indirectly benefiting from the exploitation of the North and the tale of impending species loss that the film invites spectators to mourn. The images of and voices from the region engaged in living out and debating its own future override that position, however. The elegiac tone derives in part from the documentation of the feared loss of a still-current way of life, fought for by an impassioned portion of the indigenous peoples themselves. The film documents elements of the story of a conflicted people trying to carve out their rightful place in—and in spite of—an outside-imposed economic order. In so doing, the film also bears witness to a portion of the indigenous population actively adapting to this new order, working to exploit it for the good of their people. Notably, the filmmaker interviews Stephen Kakfwi, a Sahtu Dene, the minister of what became the Department of Resources, Wildlife and Economic Development of the Northwest Territories and an advocate for northern management of the region's affairs. (He later became premier of the Northwest Territories.)

While the film attends to various positions and concerns voiced by the Aboriginal people, it keeps the anticipatory nature of its elegiac components largely centered on the fate of the tundra's nonhuman species. This focus implicitly questions whether all nonhuman life-forms can adapt to, let alone survive, the coming change. It is around the wild species, not the First Peoples of the North, that the anxieties of the film's extinction discourse swirl. These anxieties are powerfully drawn together in the film's deployment of the Eskimo curlew's extinction narrative as a central vehicle to advocate for the present and future of the North's wild species and wild spaces. In sum, the documentary stakes out a final stand on their behalf by widening the symbolic reference of the Eskimo curlew to still-extant northern species (such as the caribou) and then inviting all Canadians, but notably the dominant population of Euro-Canadians in the South, to protect the remaining wild species in the Canadian North. The filmmaker draws on the elegiac mode to rechannel viewers' shock and sadness over the extent of humans' implication in the destruction of a particular species. The film thus warns of the impending decline of other abundant species unique to the North in an implicit attempt to incite action. To underscore its focus on the land and its wild species, the film opens with images of the barrens and then brackets the beginning and ending with images of various emblematic species, linking extant migratory birds with more northerly circumscribed mammals, such as the musk ox, polar bear, and caribou. All, suggests the film, are threatened by forms of human activity unnecessary to our survival.

And Now?

With the passage of time, the film has become a historical artifact, another document recording the recurrent collision of the forces of modernity with those advocating the preservation of a fragile ecosystem and the traditional lifestyle it supports. Now yet another milestone in the cultural journey of the Eskimo curlew extinction narrative, the film, in today's context, invites us to question how persuasive the elegiac mode is in helping to mitigate human environmental damage in the North. The three texts use various elegiac strategies to instigate southern-dwelling North Americans to care about and take responsibility for harm they at least indirectly cause northern species. Regardless of the elegiac approach—whether built from a Canadian, American, or largely nonnational perspective—one is left wondering about the effectiveness of the mode itself.

The forces of capitalism, industrialization, and resource extraction serving an ever increasing, ever more demanding, and progressively encroaching human population are admittedly Goliathan, helping explain the inability of the mode to reach, let alone move, a critical mass of people from merely listening to taking action. Economic interests distract and dominate. It would be hubris to expect otherwise, influential anticipatory elegies of the likes of Rachel Carson's *Silent Spring* (1962) notwithstanding. Mining and quarrying along with oil and gas extraction together constitute the largest single sector of the economies of both the Northwest Territories, where the Eskimo curlew was mainly known to nest, and Alaska, where it is thought to have once nested also. Indeed, both Canada and the United States adhere to a largely neoliberal economic model of resource extraction, treating the North as a so-called last frontier of riches for the taking.

Are the similarities enough for attempting to redeploy across both jurisdictions an environmental symbol like the Eskimo curlew—elegiac or otherwise—on behalf of wild creatures of the North and their extensive habitats? Differences in the weight of these dominant industries in each jurisdiction and attendant national policies may, at times, need to be taken into account in order for a transnational, environmentally inflected narrative to be effective. This could mean recognizing that the preponderance of specific industries in the Northwest Territories and Alaska differ, with mining, notably diamond mining, dominating in Canada and oil and gas extraction dominating in Alaska.[58] Ranging wider in Canada's norths, other headline-catching primary industries complicate the mix, from the crude oil industry concocted from the vast stretches of bituminous sands in northern Alberta to uranium mining in northern Saskatchewan. The list goes on when the bewildering breadth of mining concerns are considered, inviting narrative strategies both locally and globally engaged.

It is out of the scope of this essay to determine whether conservation needs to be patriotic or nation-free in order to be effective; however, the findings here invite a comment. As these texts show, given the current geopolitical division of the North American Arctic, each approach has its raison d'être and will be favored depending on target audience. Ideally, though, a multinational approach would be part of any future mix. If the fate and past interventions made in the interests of a migratory bird like the Eskimo curlew still teach us anything today, it is that multinational efforts on behalf of wild species that wander widely can be salutary. It is a question of timing and the size of the human population engaging in the efforts. A case in point is the Migratory Bird Convention of 1916, a treaty between the United States and Canada that is still in effect today. The decimation of the Eskimo curlew—along with that of other indiscriminately hunted migratory birds, most famously the passenger pigeon—established the treaty's need. Though it came too late to save the Eskimo curlew, it helped other species recover.[59]

The question of timing brings me to my final point. As a mode, elegy is bound up with time, traditionally the past. In proleptic elegies, such as ecoelegies, the future is also implicated—or, as more than one critic has pointed out, the future perfect or, in some cases, the future progressive. These elegies invite prospective mourning, sometimes to unspecified future endpoints. As such, as much as they may impel current action to forestall a bad end, they may also inhibit action. They may paralyzingly wrap the end into now, heralding the absolute despair of a vague future time. In the case of the Eskimo curlew, all real hope of saving the species was already extinguished by the time Bodsworth's novel was published. The bird and its fate could only be emblematic. It had become, in effect, a ghost—a spirit mobilizing action.

The difficulties posed by environmental elegies are further complicated by the fact that in the past their language became inadvertently entangled in that of these anticipatory elegies about the vanishing of indigenous peoples, as in the case of Bodsworth's novel. This entanglement can undermine the ecoelegies' message with unintended ideological baggage. They can end up backhandedly suggesting that the population of a species is not declining (the false message behind the old premise of the "vanishing race") when in fact it is. While care can be taken to avoid these and previously noted pitfalls by writing in a more enlightened vein, Heise's suggestion to seek alternatives to the elegiac mode nevertheless warrants consideration. Perhaps there are other ways to tell the continuation of the story these texts began in order to voice concerns and incite action on behalf of imperiled northern species and their habitat spaces.

While broadly speaking the comic mode, as Heise suggests, may have a place in such efforts, the satiric strain of this approach has particular appeal, given how the stories of the Eskimo curlew and the species with which it has been attached have continued to play out. The comic mode, which as Meeker explains culminates in renewal and regeneration and emphasizes adaptation and integration, may allow for a hopeful vision of survival in a degrading world. But it also raises the question, whose survival? As some species, among them *Homo sapiens*, adapt, what of those that die? As ecosystems reconfigure and regenerate, what of those that are impoverished? Do we aspire to a North or Norths more or less biologically complex? How can regeneration be imagined? How long might it take? And in the meantime?

The period since ornithologists began predicting the extinction of the Eskimo curlew a good century ago reveals much to mourn, and much to satirize. The passage of time seems, as with an anticipatory elegy, to intensify the message, but in this case by helping to reveal the extent of human folly contributing to the fate of this bird and to the fate of too many of its fellow wild creatures. The epilogue to the current phase in the story of the Eskimo curlew and some of its cultural referents implicates southerners. The few indicators of environmental change, outlined below, imply that those of us in the lands to the south have been bent on continuing, blithely it would seem, on our consumptive path in and perhaps more accurately *of* the North, regardless of scientists' warnings and the decline of species unique to the region. Industry—driven largely by southern needs and wants—continues to extract, disrupt, and move on. While government and industry coffers may fill, ecoledgers record losses, many immense. Even the criteria that need to be met before definitively declaring the emblematic bird extinct seem overly cautious and exhaustive in the face of all scientific probability. Could one say another form of human evasion? Foolish blind hope? And what of those species that are expected to move in as temperatures rise? What does their eventual advent say about our commitment to biodiversity and the unique features of what we generically called "the North"? Is this mere southern nostalgia or a call for restorative action?

A satiric rendition of this continuing, complex story may be no more able to impel readers or viewers of future texts to change the course of events in the North or Norths than ecoelegies have. Perhaps, though, it would invite other ways to envision them, other points of emphasis—not on endpoints, not on new beginnings—but on the absurd reality bare facts reveal: our inability to forestall what was foretold, symptoms of our complicit entrapment in economic, political, and social systems of our own making. Perhaps, too, it will offer another emotional path to action: a catalytic—if we like a bitingly caustic—way of grappling with the unfolding situation. Indeed, in the end, we may discover that satire will not suffice either, that fluctuating between modes, combining modes, and inventing new ones will be best for expressing our response to the shifting ground ahead: one that will move us to mourn, to rail, to fight, to restore, to adapt, to mourn again, to query, to rethink, and to act. Dare we hope that modal diversity may help beget biodiversity? The building blocks are here for the retelling.

Epilogue

More than fifty years have passed since the last confirmed observation of the Eskimo curlew in 1963, the requisite period for declaring a species extinct. The IUCN Red List of Threatened Species deems it critically endangered (possibly extinct), pending the cessation of occasional unconfirmed sightings and the completion of surveys of potential nesting areas.[60]

The populations of many of the world's arctic-breeding migratory shorebirds are now in serious decline. Ditto for the populations of many birds migrating out of the boreal forest. The Eskimo curlew was the proverbial canary in the mineshaft. In 2013, the Arctic Council, through its Conservation of Arctic Flora and Fauna working group, initiated the Arctic Migratory Birds Initiative in the hopes of inducing countries along the birds' flyways—in the North *and* in the South—to turn the tide.

The population of the Bathurst caribou herd, for which the fate of the Eskimo curlew became a foreboding portent in *The Barrens Quest*, dropped 93 percent "from a peak of 472,000 in 1986 to 32,000" in 2010,[61] devastatingly fulfilling the predictions of Anne Gunn, one of the biologists interviewed in the documentary film in 1996.

In 2013, BHP completed the sale of its controlling stake in the mine at issue in the film: the Ekati Diamond Mine, Canada's first. At the time of initial sale in 2012, the mine was thought to be seven years shy of its expected life, barring the discovery of further reserves.[62]

Today the majority of the circumpolar caribou and reindeer herds for which data are available are in decline, including the Porcupine herd that migrates across the American and Canadian North. Some, like the Peary herd, face annihilation. While the specific reasons for this phenomenon differ from those that decimated the Eskimo curlew, root causes can ultimately be attributed directly or indirectly to the effects of human activity on the ecosystem of which the caribou is part: weather disruption, habitat change, and disturbances caused by mineral, oil, and gas extraction in the North.[63]

Although in the wake of climate warming wildlife biologists predict the continued decline of some northern species, notably the caribou and the polar bear, they anticipate that other species nondependent on a cold climate may thrive, such as white-tailed deer and raccoons.[64] More adaptable, these particular species are not unique to the region. Will those that finally succumb be mainly, like the Eskimo curlew, the wild creatures that embody—as some said the Eskimo curlew did—"the spirit of 'the North'"? What, then, will the new North(s) be?

Who to say? Who to care? The verb—ours.

Acknowledgments

My thanks to Sarah Keller for her research assistance.

Endnotes

1. Ursula K. Heise, "Lost Dogs, Last Birds, and Listed Species: Cultures of Extinction," *Configurations* 18, no. 1–2 (Winter 2010): 49–72. First quotation, p. 60. Second quotation, p. 50. Third quotation, p. 72. Also see pp. 49 and 52–60.

2. Another challenge is the definition of "North" itself. For seminal work in the Canadian context, see Sherrill E. Grace's *Canada and the Idea of North* (Montreal and Kingston: McGill-Queen's University Press, 2001). She argues that "North . . . is a human construct" (15). While for her it is "multiple, shifting and elastic . . . a *process*" (16), it is also "historically lived and changing *and* spatially configured, even if . . . *very* vaguely and imprecisely" (22). Also see the work of geographer Louis-Edmond Hamelin for his concept of "nordicity" along with his terminology for its study. His gradients of the North (Middle North, Far North, Extreme North), measured by physical and sociocultural variables, help in the conceptualization of a vast and complex space. He nevertheless also acknowledges the notion of multiple Norths, at once shifting and mentally construed. His work in translation includes *Canadian Nordicity: It's Your North, Too*, trans. William Barr (Montreal: Harvest House, 1979) and *The Canadian North and Its Conceptual Referents* (Ottawa: Canadian Studies Directorate, 1988). A list of two hundred terms and some explanations are found in his text *Discours du Nord* (Quebec: GÉTIC, Collection Recherche 35, Université Laval,

2002). Building on this earlier work, Daniel Chartier has been directing the research center Imaginaire du Nord at the Université du Québec à Montréal since 2003. An international team of researchers has been doing comparative cultural studies on the representation of the North and winter in texts from Quebec and elsewhere, grappling with questions of epistemology and methodology.

3. For a general, nonregional overview, see David W. Orr, "Baggage: The Case for Climate Mitigation," *Conservation Biology* 23, no. 4 (2009): 790–793, especially page 793 on which he affirms, "No matter what we do to adapt, . . . we will lose many species." On the degree of the threat of extinction in Canada, see Laura E. Coristine and Jeremy T. Kerr, "Habitat Loss, Climate Change, and Emerging Conservation Challenges in Canada," *Canadian Journal of Zoology* 89 (2011): 435–451. For a list of species at risk in Canada, see the Species at Risk Public Registry at http://www.sararegistry.gc.ca/.

4. Observes Joseph W. Meeker, "The tragic man takes his conflict seriously, and feels compelled to affirm his mastery and his greatness in the face of his own destruction." "The Comic Mode," in *The Ecocriticism Reader: Landmarks in Literary Ecology*, ed. Cheryll Glotfelty and Harold Fromm (Athens: University of Georgia Press, 1996), 157. The quotation cited in the body of the essay is from p. 158.

5. In this essay I am following Heise's lead and loosely separating "elegiac and tragic story templates" (op. cit. 49). Not all critics do this. For example, Northrop Frye, in his theory of modes, subsumes elegiac literature under one of the categories he proposes for the tragic fictional mode (*Anatomy of Criticism*, [1975; repr., Princeton: Princeton University Press, 1973, 36). Classifying his modes according to the "hero's power of action" (33), he distinguishes, for instance, between high and low mimetic tragedy, the former dealing with the fall of a leader, the latter with the exclusion of "one of us" (34) from a social group, which could take the form of "the death of an animal" (39). It is out of the scope of this essay to consider how Frye's modal categories for tragedy might open up Meeker's narrowly applied paradigm of classical tragedy in making relationships between ecology and literature, but it would seem worth pursuing.

6. Chris Baldick, "Elegy," in *Oxford Dictionary of Literary Terms* (Oxford: Oxford University Press, 2008), 105.

7. Timothy Morton, "The Dark Ecology of Elegy," in *The Oxford Handbook of the Elegy*, ed. Karen Weisman (Oxford: Oxford University Press, 2010), 251–271. See 254.

8. Ibid.

9. Bonnie Costello, "Fresh Woods: Elegy and Ecology Among the Ruins," in Weisman, *The Oxford Handbook of the Elegy*, 324–342. See 330.

10. Patrick Brantlinger, *Taming Cannibals: Race and the Victorians* (Ithaca: Cornell University Press, 2011), 2. See also Patrick Brantlinger, *Dark Vanishings: Discourse on the Extinction of Primitive Races, 1800–1930* (Ithaca: Cornell University Press, 2003), 3–5.

11. George Archibald, "Following the Cranes to the Arctic," in *Arctic Voices: Resistance at the Tipping Point*, ed. Subhankar Banaerjee (New York: Seven Stories Press, 2012).

12. Margaret Stroebe and Henk Schut, "The Dual Process Model of Coping with Bereavement: A Decade On," *Omega—Journal of Death and Dying* 61, no. 4 (2010): 273–289.

13. See, for example, Mishka Lysack, "Emotion, Ethics, and Fostering Committed Environmental Citizenship," in *Environmental Social Work*, ed. Mel Gray, John Coates, and Tiani Hetherington, 231–245 (London: Routledge, 2012). See 238.

14. R. Clifton Spargo, *The Ethics of Mourning: Grief and Responsibility in Elegiac Literature* (Baltimore: The Johns Hopkins University Press, 2004). See his introduction pp. 1–13, notably p. 6.

15. Costello, "Fresh Woods," 330.

16. This essay refers to the 2010 edition: Fred Bodsworth, *Last of the Curlews* (1955; repr., Toronto: McClelland and Stewart, 2010). Bodsworth's fictional account of a nearly extinct bird is predated by a nonfictional account of the quest for a nearly lost bird: James T. Tanner, *The Ivory-Billed Woodpecker* (1942; repr., Mineola, NY: Dover, 2003).

17. John W. Fitzpatrick, "The American Ornithologists' Union and Bird Conservation: Recommitment to the Revolution," in *Bird Conservation, Implementation and Integration in the Americas: Proceedings of the Third International Partners in Flight Conference*, ed. John Ralph and Terrell D. Rich (USDA, Forest Service, Gen. Tech. Rep. PSW-GTR-191, 2005), 46. Also see Janet Foster, *Working for Wildlife: The Beginning of Preservation in Canada* (Toronto: University of Toronto Press, 1978), 120–148.

18. Although the novel crops up in scientific articles about the Eskimo curlew and although literary scholars mention it in passing, notably in studies on animals or birds in Canadian literature, it tends to be made short shrift of in cultural studies on the North. For example, Margaret Atwood uses it in her book *Survival* (Toronto: Anansi Press, 1972; p. 76) to illustrate her thesis that sacrificed animal figures symbolize Canada's sense of neocolonial dominance. It is barely mentioned, however, in Sherrill Grace's *Canada and the Idea of North* (Montreal: McGill-Queen's University Press, 2007, p. 184), which focuses on what Grace qualifies as "serious fiction" (xiv) and is ignored in Renée Hulan's *Northern Experience and the Myths of Canadian Culture* (Montreal and Kingston: McGill-Queen's University Press, 2001). Allison Mitcham notes that in the novel the senseless killing of the bird is used to advance a value judgment: to contrast sports-based hunting by humans with needs-based killing for food by wild creatures. Mitcham, *The Northern Imagination* (Moonbeam, ON: Penumbra Press, 1983), 26. Other scholarship will be noted in conjunction with specific parts of the analysis.

19. *The Last of the Curlews* (Hanna-Barbera Studios, 1972).

20. *The Barrens Quest*, directed by Peter Blow (Bar Harbour Films/Lindum Films/National Film Board of Canada/Canadian Broadcasting Corporation, 1997). See https://www.nfb.ca/film/the_barrens_quest.

21. Since the 1970s the North American public had been reminded of the impact of development in the North. In the late 1960s and early 1970s, environmentalists in the United States raised concerns about pipeline development in Alaska, touted as the last American wilderness. Two inquiries held in the 1970s into pipeline development in the Canadian North, one in the Northwest Territories (the Mackenzie Valley Pipeline Inquiry, headed by Justice Thomas Berger) and one in the Yukon (the Alaska Highway Pipeline Inquiry, headed by Kenneth M. Lysyk) forcefully brought northern issues into the public eye in Canada. See Shelagh D. Grant, *Polar Imperative: A History of Arctic Sovereignty in North America* (Vancouver: Douglas & McIntyre, 2010), 360–361 and 364–366.

22. Heise, "Lost Dogs, Last Birds, and Listed Species," 62.

23. Brantlinger, *Dark Vanishings*, 2.

24. David A. Kirk and Jennie L. Pearce, *COSEWIC Assessment and Status Report on the Eskimo Curlew* Numenius borealis *in Canada* (Canadian Wildlife Service, Environment Canada, 2006). Scientists now believe that the Eskimo curlew's population at the time of European contact was more likely in the several hundreds of thousands (14–15).

25. The lack of article "the" at the beginning of the title keeps the status of the curlews followed in the text ambiguous.

26. Scientists evoke three main factors in its demise: massive overhunting, habitat loss, and the extinction of one of its food staples in the American West. See Kirk and Pearce, *COSEWIC Assessment*, 18–20, and "Eskimo Curlew: Three Strikes in the Wink of an Eye," The Cornell Lab of Ornithology, http://www.birds.cornell.edu/AllAboutBirds/conservation/extinctions/eskimo_curlew.

27. The farmer also evokes the latent image of the hunter, since he shoots the bird. Two lethal forces that conspired against the Eskimo curlew—venatic and agricultural—are thus combined in this one act.

28. These two quotations appear in Bodsworth, *Last of the Curlews*, 27.

29. Rather, it is described in terms of land formations, rocks, and a "river's S twist" (ibid., 6).

30. That the curlew pair's final destination lies in Canada is suggested by the single phrase "[i]n Canada far to the north," which introduces the description of spring coming to the Arctic while the curlews prepare to cross the Gulf of Mexico (Bodsworth, *Last of the Curlews*, 96).

31. Lee Frew, "A Kinship with Otherness: Settler Subjectivity and the Image of the Wild Animal in Canadian Fiction in English" (PhD diss., Toronto: York University, 2011), 212.

32. J. B. Gollop, T. W. Barry, and E. H. Iversen state, "There is no concrete evidence for breeding in Alaska." "Eskimo Curlew: A Vanishing Species?" subsection: "Fall Migration in Northern North America." See also Map 3. Saskatchewan Natural History Society. Special Publication No. 17, Regina, Saskatchewan, 1986; http://www.npwrc.usgs.gov/resource/birds/curlew/.

33. These places include Labrador, New England, Nantucket, Texas, Nebraska, the Dakotas, and the Canadian prairie.

34. The one mention of the land's original inhabitants, the "Eskimos" (1), depicts them as hunters of the curlew—but only for food, not for wanton sport and profit as the hunters to the south. They also serve as the repository of ecological cultural history. With living memory of the great flocks of Eskimo curlew having been lost, only legend, observes the narrative, remains.

35. This is the translation given in *The Barrens Quest*; it appears to be based on a translation of the Latin *numen minis*: "divine will; divine power, majesty; deity, god." B. C. Taylor and K. E. Prentice, *Selected Latin Readings* (J.M. Dent & Sons, 1962), 370. However, several scientific publications give alternative translations based on Latinized Greek roots. For a review of these translations, see Gollop, Barry, and Iversen, "Eskimo Curlew."

36. Scientific reports distinguish between known nesting sites (in the Northwest Territories) and sightings in flight farther afield, as in Alaska.

37. Bodsworth, *Last of the Curlews*, 66.

38. Ibid., 109.

39. The novel uses the term "home" (68). Lee Frew reads the migrating curlew pair as "tourists" contrasted with "two images of the exogene, the South American llama and Indian herder" (213).

40. See pp. 32 and 33 of the novel. Other excerpts from reports note the shocking drop in numbers between 1882 and 1894 (52).

41. See p. 2 of the novel.

42. Frew, "A Kinship with Otherness," 216–217.

43. For examples of this strand of discourse, see J. B. MacKinnon, *The Once and Future World* (Toronto: Random House Canada, 2013), 28–35.

44. See, for example, Kenneth Coates, *Canada's Colonies: A History of the Yukon and the Northwest Territories* (Toronto: James Lorimer, 1985).

45. Brantlinger, *Dark Vanishings*, 6.

46. See Ted Sennett, William Hanna, and Joseph Barbera, *The Art of Hanna Barbera: Fifty Years of Creativity* (New York: Viking Studio Books, 1989), 210, 213;

Joseph Barbera, *My Life in 'Toons: From Flatbush to Bedrock in Under a Century* (Atlanta: Turner Publishing, 1994), 192; and Iwao Takamoto, *Iwao Takamoto: My Life with a Thousand Characters* (Jackson: University Press of Mississippi, 2009), 107–108. Takamoto adds that the animation director Nick Nichols (Charles A. Nichols), who also had once worked for Disney, was not adequately remembered for his part in the production.

47. Barbera, *My Life in 'Toons*, 192. Indeed, the film showcases Hanna-Barbera's move to a fuller animation style—somewhere between limited and full animation—that the studio began developing for animated features and specials in which it began dabbling in the late 1960s. See Sennett, Hanna and Barbera, *The Art of Hanna Barbera,* 187. The style was to enrich the backdrop of other Hanna-Barbera films featuring nonhuman characters, for example, *Charlotte's Web* (1973), and flesh out the natural world of human characters in films like *Heidi's Song* (1982).

48. While fondly remembered by now-adult viewers in numerous YouTube posts (2009–2014) on the uploaded version of the animated film, others balked at its preachy tone. See "ABC Afterschool Specials," in *The Encyclopedia of Guilty Pleasures: 1,001 Things You Hate to Love* (Philadelphia: Quirk Books, 2004), 9–10. This oppressive moral didacticism situates the film within the long tradition of morally didactic fiction for children, a tradition that reasserted itself in the 1970s when authors of children's literature began vigorously embracing contemporary environmental issues after the publication of Dr. Seuss's *The Lorax* in 1971. See Carolyn Sigler, "Wonderland to Wasteland: Toward Historicizing Environmental Activism in Children's Literature," *Children's Literature Quarterly Association* 19, no. 4 (Winter 1994): 148–153.

49. See Deidre M. Pike, *Enviro-Toons: Green Themes in Animated Cinema and Television* (Jefferson: North Carolina: McFarland & Co., 2012), 12–13, and Robin L. Murray and Joseph K. Heumann, *That's All Folks? Ecocritical Readings of American Animated Features* (Lincoln: University of Nebraska Press, 2011), 1–2, for their discussions of Jaime J. Weinman's neologism "enviro-toon." See Jaime J. Weinman, "Things That Suck: The Smoggies," "Something Old, Something New: Thoughts on Popular Culture and Unpopular Culture," Sept. 1, 2004, and "A Good Enviro-Toon," Sept. 6, 2004 (blog), http://zvbxrpl.blogspot.ca.

50. With its strong indictment of the hunting culture, the animated film joins the tradition of Disney's *Bambi.* See, for example, Ralph H. Lutts, "The Trouble with Bambi: Walt Disney's *Bambi* and the American Vision of Nature," *Forest & Conservation History* 36, no. 4 (October 1992): 160–171. Unlike *Bambi*, which does not show the actual shooting of Bambi's mother, the Hanna-Barbera production shows the shooting of the female mate, her subsequent struggle to fly, and her ultimate death.

51. If YouTube and IMDb postings on the film are any indication, the finality of the closing line marked many children who originally saw the film. Marine biologist and

writer Eva Saulitis testifies to the film's life-determining effects in her prologue to *Into the Great Silence: A Memory of Discovery and Loss among Vanishing Orcas* (Boston: Beacon Press, 2013) in which she bears witness to the last of the orcas off the Alaska coast. Originally from New York, Saulitis ventured North as a young woman and dedicated her life to the study of the whales.

52. Linda Campbell, *Endangered and Threatened Animals of Texas: Their Life History and Management* (Austin: Texas Parks and Wildlife Press, 1995, 1–3; revised and approved by USFWS 2003, p. 59 in the electronic version). Later in the entry on the Eskimo curlew, the write-up's language is a little less definitive but still relatively certain, stating the species "probably also nested in the Alaskan tundra" (p. 60 in the electronic version), http://www.tpwd.state.tx.us/publications/pwdpubs/media/pwd_bk_w7000_0013.pdf. By contrast, the report by Alaska wildlife officials claims that the Eskimo curlew "historically nested in tundra in the Northwest Territories, presumably adjacent Nunavut, and possibly in Alaska." *Eskimo Curlew (Numenius borealis) 5-Year Review: Summary and Evaluation,* U.S. Fish and Wildlife Service, Fairbanks, Alaska, August 31, 2001, 3.

53. Heise, "Lost Dogs, Last Birds, and Listed Species," 69.

54. The Eskimo curlew population collapsed between 1870 and 1890, prior to significant settlement in the Canadian West, which only began in 1896. The environmental changes this settlement wrought would have affected only the last vestiges of this particular species. For a discussion of the relative effects of hunting and agricultural development in Canada and the United States on migratory birds during this period, see Janet Green, "The Federal Government and Migratory Birds: The Beginning of a Protective Policy," *Canadian Historical Association: Historical Papers* 11, no. 1 (1976): 207–227, especially p. 210, and Dan Gottesman, "Native Hunting and the Migratory Birds Convention Act: Historical, Political, and Ideological Perspectives," *Journal of Canadian Studies* 18 (Fall 1983): 67–89, especially pp. 69–70.

55. One calls to mind other well-known cases of the extinction of northern species, such as the auk, which ranged across the North Atlantic.

56. Given this history, it would be disingenuous to argue that this outreaching rhetorical tactic militates against the pitfalls of racially or ethnically exclusive bioregional discourse, since, as Jenny Kerber has documented, this exclusionary type of discourse has typically been motivated by xenophobic or white supremacist ideologies. See Jenny Kerber, "Nature Trafficking: Writing and Environment in the Western Canada–U.S. Borderlands," in *Greening the Maple: Canadian Ecocriticism in Context*, ed. Ella Soper and Nicholas Bradely (Calgary: University of Calgary Press, 2013), 199–225. See pp. 207–210.

57. Renato Rosaldo, "Imperialist Nostalgia," *Representations* 26 (Spring 1989): 107–122.

58. "Northern Economic Diversification Index," Canadian Northern Development Agency, http://www.cannor.gc.ca/eng/1388762115125/1388762170542, accessed June 11, 2014. In 2012, mining, quarrying, and oil and gas extraction accounted for 28 percent of the NWT's GDP. In 2005, these sectors accounted for 27.2 percent of gross regional product (GRP) in arctic Canada and 27.9 of GRP in Alaska; however, oil and gas extraction dominated in Alaska (23.9%) while mining dominated in arctic Canada (19.5%). Solveig Glomsrød et al., "Arctic Economies within the Arctic Nations," in *The Economy of the North 2008*, ed. Solveig Glomsrød and Iulie Aslaksen (Oslo-Konsvinger: Statistics Norway, 2010), 37–67. Neoliberal model: 16–17; Alaska (by industry): Table 4.1, p. 38, Arctic Canada (by industry): Table 4.2, p. 42. http://www.ssb.no/a/english/publikasjoner/pdf/sa112_en/sa112_en.pdf, retrieved June 11, 2014.

59. It also raised issues on how to allow for indigenous people's hunting, the central concern in Fielding's essay (this volume). For all the Migratory Bird Treaty's positive effect on bird populations, the tapestry of laws and regulations dealing with biodiversity in Canada and the continental United States have been found ineffective as a whole in the face of "the ultimate drivers of biodiversity loss." Reed F. Noss, et al., "Priorities for Improving the Scientific Foundation of Conservation Policy in North America," *Conservation Biology* 23, no. 4 (2009): 825–833; citation from p. 826.

60. A specimen was culled in the Barbados in 1963. For a discussion of sightings since, see Kirk and Pearce, *COSEWIC Assessment and Status Report*, 15–16. For sightings prior to 1963, see Gollop, Barry, and Iversen, "Eskimo Curlew."

61. Ed Struzik, "A Troubling Decline in the Caribou Herds of the Arctic," *Yale Environment* 360 (blog), Yale School of Forestry and Environmental Studies, Sept. 23, 2010, http://e360.yale.edu/feature/a_troubling_decline_in_the_caribou_herds_of_the_arctic/2321/.

62. See Pav Jordan, "BHP Billiton Breaks Its Diamond Engagement," *Globe and Mail*, Nov. 13, 2012, http://www.theglobeandmail.com/report-on-business/industry-news/energy-and-resources/bhp-billiton-breaks-its-diamond-engagement/article5227127/ and "Dominion Diamond Wins Final Approvals to Acquire Ekati Mine," *Financial Post*, April 1, 2013, http://business.financialpost.com/2013/04/01/dominion-diamond-ekati/.

63. Liv Solveig Vors and Mark Stephen Boyce, "Global Declines of Caribou and Reindeer," *Global Change Biology* 15, no. 1 (2009): 2626–2633, p. 2626.

64. Mark S. Boyce, et al., "Harvest Models for Changing Environments," in *Wildlife Conservation in a Changing Climate*, ed. Jedediah F. Brodie, Eric S. Post, and Daniel F. Doak (Chicago: University of Chicago Press, 2012), 293–306, p. 293.

Bibliography

Archibald, George. "Following the Cranes to the Arctic." In *Arctic Voices: Resistance at the Tipping Point*, edited by Subhankar Banaerjee. New York: Seven Stories Press, 2012.

Atwood, Margaret. *Survival*. Toronto: Anansi Press, 1972.

Baldick, Chris. "Elegy." In *Oxford Dictionary of Literary Terms*. Oxford: Oxford University Press, 2008.

Barbera, Joseph. *My Life in 'Toons: From Flatbush to Bedrock in Under a Century*. Atlanta: Turner Publishing, 1994.

The Barrens Quest, directed by Peter Blow. Bar Harbour Films/Lindum Films/National Film Board of Canada/Canadian Broadcasting Corporation, 1997.

Bodsworth, Fred. *Last of the Curlews*. 1955; repr., Toronto: McClelland and Stewart, 2010.

Boyce, Mark S., et al. "Harvest Models for Changing Environments." In *Wildlife Conservation in a Changing Climate*, edited by Jedediah F. Brodie, Eric S. Post, and Daniel F. Doak, 293–306. Chicago: University of Chicago Press, 2012.

Brantlinger, Patrick. *Taming Cannibals: Race and the Victorians*. Ithaca: Cornell University Press, 2011.

———. *Dark Vanishings: Discourse on the Extinction of Primitive Races, 1800–1930*. Ithaca: Cornell University Press, 2003.

Campbell, Linda. *Endangered and Threatened Animals of Texas: Their Life History and Management*. Austin: Texas Parks and Wildlife Press, 1995.

Canadian Northern Development Agency. "Northern Economic Diversification Index." Acccessed June 11, 2014. http://www.cannor.gc.ca/eng/1388762115125/1388762170542.

Coates, Kenneth. *Canada's Colonies: A History of the Yukon and the Northwest Territories*. Toronto: James Lorimer, 1985.

Coristine, Laura E., and Jeremy T. Kerr. "Habitat Loss, Climate Change, and Emerging Conservation Challenges in Canada." *Canadian Journal of Zoology* 89 (2011): 435–451.

Costello, Bonnie. "Fresh Woods: Elegy and Ecology Among the Ruins." In *The Oxford Handbook of the Elegy*, edited by Karen Weisman, 324–342. Oxford: Oxford University Press, 2010.

Eskimo Curlew (Numenius borealis) 5-Year Review: Summary and Evaluation, U.S. Fish and Wildlife Service, Fairbanks, Alaska, August 31, 2001, 3.

Fitzpatrick, John W. "The American Ornithologists' Union and Bird Conservation: Recommitment to the Revolution." In *Bird Conservation, Implementation and Integration in the Americas: Proceedings of the Third International Partners in Flight*

Conference, edited by John Ralph and Terrell D. Rich. USDA, Forest Service, Gen. Tech. Rep. PSW-GTR-191, 2005.

Foster, Janet. *Working for Wildlife: The Beginning of Preservation in Canada*. Toronto: University of Toronto Press, 1978.

Frew, Lee. "A Kinship with Otherness: Settler Subjectivity and the Image of the Wild Animal in Canadian Fiction in English." PhD diss., Toronto: York University, 2011.

Frye, Northrop. *Anatomy of Criticism*. 1975; repr., Princeton: Princeton University Press, 1973.

Glomsrød, Solveig, et al. "Arctic Economies within the Arctic Nations." In *The Economy of the North 2008*, edited by Solveig Glomsrød and Iulie Aslaksen, 37–67. Oslo-Konsvinger: Statistics Norway, 2010.

Gollop, J. B., T. W. Barry, and E. H. Iversen. "Eskimo Curlew: A Vanishing Species?" Saskatchewan Natural History Society. Special Publication No. 17, Regina, Saskatchewan, 1986; http://www.npwrc.usgs.gov/resource/birds/curlew/.

Grace, Sherrill E. *Canada and the Idea of North*. Montreal and Kingston: McGill-Queen's University Press, 2001.

Grant, Shelagh D. *Polar Imperative: A History of Arctic Sovereignty in North America*. Vancouver: Douglas & McIntyre, 2010.

Green, Janet. "The Federal Government and Migratory Birds: The Beginning of a Protective Policy." *Canadian Historical Association: Historical Papers* 11, no. 1 (1976): 207–227,

Gottesman, Dan. "Native Hunting and the Migratory Birds Convention Act: Historical, Political, and Ideological Perspectives." *Journal of Canadian Studies* 18 (Fall 1983): 67–89.

Hamelin, Louis-Edmond. *Canadian Nordicity: It's Your North, Too*. Trans. William Barr. Montreal: Harvest House, 1979.

———. *The Canadian North and Its Conceptual Referents*. Ottawa: Canadian Studies Directorate, 1988.

———. *Discours du Nord*. Quebec: GÉTIC, Collection Recherche 35, Université Laval, 2002.

Heise, Ursula K. "Lost Dogs, Last Birds, and Listed Species: Cultures of Extinction." *Configurations* 18, no. 1–2 (Winter 2010): 49–72.

Hulan, Renée. *Northern Experience and the Myths of Canadian Culture*. Montreal and Kingston: McGill-Queen's University Press, 2001.

Jordan, Pav. "BHP Billiton Breaks Its Diamond Engagement." *Globe and Mail,* Nov. 13, 2012, http://www.theglobeandmail.com/report-on-business/industry-news/energy-and-resources/bhp-billiton-breaks-its-diamond-engagement/article227127/.

———. "Dominion Diamond Wins Final Approvals to Acquire Ekati Mine." *Financial Post*, April 1, 2013, http://business.financialpost.com/2013/04/01/dominion-diamond-ekati/.

Kerber, Jenny. "Nature Trafficking: Writing and Environment in the Western Canada–U.S. Borderlands." In *Greening the Maple: Canadian Ecocriticism in Context*, edited by Ella Soper and Nicholas Bradely, 199–225. Calgary: University of Calgary Press, 2013.

Kirk, David A., and Jennie L. Pearce. *COSEWIC Assessment and Status Report on the Eskimo Curlew* Numenius borealis *in Canada*. Canadian Wildlife Service, Environment Canada, 2006.

The Last of the Curlews. Hanna-Barbera Studios, 1972.

Lutts, Ralph H. "The Trouble with Bambi: Walt Disney's *Bambi* and the American Vision of Nature." *Forest & Conservation History* 36, no. 4 (October 1992): 160–171.

Lysack, Mishka. "Emotion, Ethics, and Fostering Committed Environmental Citizenship." In *Environmental Social Work*, edited by Mel Gray, John Coates, and Tiani Hetherington, 231–245. London: Routledge, 2012.

MacKinnon, J. B. *The Once and Future World*. Toronto: Random House Canada, 2013.

Meeker, Joseph W. "The Comic Mode." In *The Ecocriticism Reader: Landmarks in Literary Ecology*, edited by Cheryll Glotfelty and Harold Fromm. Athens: University of Georgia Press, 1996.

Mitcham, Allison. *The Northern Imagination*. Moonbeam, ON: Penumbra Press, 1983.

Morton, Timothy, "The Dark Ecology of Elegy." In *The Oxford Handbook of the Elegy*, edited by Karen Weisman, 251–271. Oxford: Oxford University Press, 2010.

Murray, Robin L., and Joseph K. Heumann. *That's All Folks? Ecocritical Readings of American Animated Features*. Lincoln: University of Nebraska Press, 2011.

Noss, Reed F., et al. "Priorities for Improving the Scientific Foundation of Conservation Policy in North America." *Conservation Biology* 23, no. 4 (2009): 825–833.

Orr, David W. "Baggage: The Case for Climate Mitigation." *Conservation Biology* 23, no. 4 (2009): 790–793.

Pike, Deidre M. *Enviro-Toons: Green Themes in Animated Cinema and Television*. Jefferson, NC: McFarland & Co., 2012.

Rosaldo, Renato. "Imperialist Nostalgia." *Representations* 26 (Spring 1989): 107–122.

Saulitis, Eva. *Into the Great Silence: A Memory of Discovery and Loss among Vanishing Orcas*. Boston: Beacon Press, 2013.

Sennett, Ted, William Hanna, and Joseph Barbera. *The Art of Hanna-Barbera: Fifty Years of Creativity*. New York: Viking Studio Book, 1989.

Sigler, Carolyn. "Wonderland to Wasteland: Toward Historicizing Environmental Activism in Children's Literature." *Children's Literature Association Quarterly* 19, no. 4 (Winter 1994): 148–153.

Spargo, R. Clifton. *The Ethics of Mourning: Grief and Responsibility in Elegiac Literature.* Baltimore: The Johns Hopkins University Press, 2004.

Stroebe, Margaret, and Henk Schut. "The Dual Process Model of Coping with Bereavement: A Decade On." *Omega—Journal of Death and Dying* 61, no. 4 (2010): 273–289.

Struzik, Ed. "A Troubling Decline in the Caribou Herds of the Arctic." *Yale Environment 360* (blog), Yale School of Forestry and Environmental Studies, Sept. 23, 2010, http://e360.yale.edu/feature/a_troubling_decline_in_the_caribou_herds_of_the_arctic/2321/.

Takamoto, Iwao. *Iwao Takamoto: My Life with a Thousand Characters.* Jackson: University Press of Mississippi, 2009.

Tanner, James T. *The Ivory-Billed Woodpecker.* 1942; repr., Mineola, NY: Dover, 2003.

Vors, Liv Solveig, and Mark Stephen Boyce. "Global Declines of Caribou and Reindeer." *Global Change Biology* 15, no. 1 (2009): 2626–2633.

Weinman, Jaime J. "Things That Suck: The Smoggies." "Something Old, Something New: Thoughts on Popular Culture and Unpopular Culture," Sept. 1, 2004, and "A Good Enviro-Toon," Sept. 6, 2004 (blog) http://zvbxrpl.blogspot.ca.

2

Ravens' World

Ecoelegy and Beyond in a Changing North

will elliott / UNIVERSITY OF ALASKA SOUTHEAST

Elegizing Alaska as a Lost Frontier

Literary and cinematic elegies abound for northern landscapes and lifeways imagined lost to modernization. The titular resonances of recent Alaska nonfiction—*Leaving Alaska, The Final Frontiersman, The Last Settlers,* or *Disappearance, a Map*—register a wider elegiac trend representing the North as a place of disappearance, or a place itself disappearing.[1] As Eric Heyne observes, many of Alaska literature's most canonical texts, such as the poet John Haines's memoir *The Stars, the Snow, the Fire,* "tend to look backward, and their tone is predominantly nostalgic or elegiac."[2] Such melancholy narratives not only still subtend much of American environmental art but often shape scholarly discourse on literature and environment. Yet in recent years, Alaska writers are increasingly looking beyond the elegiac mode. Among them, Heyne names Nancy Lord, whose short fiction collection *The Man Who Swam with Beavers* (2001) engages with non-Western narrative traditions and locales and adopts an often comic, ironic, or satiric tone.[3] Here I focus on a more recent Alaska writer, Seth Kantner, whose work is helping to shift the focus of contemporary Alaska literature away from elegizing what is lost to articulating how contemporary northerners, human and nonhuman, are going on living in the midst of accelerating cultural and environmental change.

Heyne's call for forward-looking northern narratives echoes a prior appeal by William Hunt, writing a generation earlier.[4] In a 1978 essay reviewing literary treatments of Alaska's transformation by oil development, Hunt caricatures "lost frontier" writers as those disheveled figures one finds in comic strips with signs that warn The End Is Near:

> Recently the writing trend is to lose the Alaskan frontier as an incident to finding it. It must frustrate Americans who have not been too keenly aware of Alaska to learn suddenly that the 'last frontier' is lost. It is not lost, last, or even a frontier. Let's settle that, and save the wilderness concept for a separate consideration. (22)

Objecting to finalizing essentialisms like "last" and "lost," Hunt insists instead on "the opportunity to weigh specific land proposals against competing proposals without entanglement by wilderness visions" (ibid.). Yet perhaps we can push Hunt's critique further, to consider how the stakes of elegizing Alaska encompass the broader question of Alaska's relationship to late-capitalist modernity.

As Elspeth Tulloch observes in chapter 1, apocalyptic environmental rhetoric runs the risk of positioning itself as a self-fulfilling prophecy. Timothy Morton identifies such prophecies' implicitly imperial logic when he asks:

> What if capitalism itself relied on fantasies of apocalypse in order to keep reproducing itself? What if Nature as such, the idea of a radical outside of the social system, was a capitalist fantasy, even *the* capitalist fantasy . . . proclaiming that history is at an end, with the capitalist form as its final destination.[5]

Figured as an exceptional space in the American environmental imagination, Alaska functions as the imagined radical outside that Morton describes—a last frontier at the top of the world whose very lastness proclaims "that history is at an end."[6] Consequently, although elegiac representations of Alaska may be effective in the short term at mobilizing audiences in response to specific threats like arctic warming, in the long term they also imply the futility of those responses, the finality of late-capitalist modernity, and the terminality of everything else.

This implicit fatalism is evident in tourist promotions, such as a recent cruise ship publication urging "if you want to see the wild side of Alaska, the most compelling reason to go now is that soon it may all be gone." This loss, the article continues, is not merely a matter of "global warming and other environmental changes" but more immediately of troublesome "new laws to protect animals from cruise ships."[7] By conflating the consumption of Alaska with the state itself, the article presents both climate change and conservation as "threats to Alaska's marquee natural attractions." In this limited-time-only landscape, resources and representations must be devoured while supplies last. Not only, then, does elegy hold little relation to lived Alaska futures, in fact it also imperils those futures by rationalizing, and in fact incentivizing, greater human impacts.

In what follows, I begin with Grant Sims's memoir, *Leaving Alaska*, which details Sims's disillusionment with Alaska's Last Frontier mythos in the wake of the 1989 *Exxon Valdez* oil spill in Prince William Sound. Critical perspectives on the politics of mourning and melancholia have addressed the consequences of Sims's affective stance in broader terms. I examine one of these formulations, Dominic Fox's notion of "the cold world" vis-à-vis that other cold world of arctic Alaska. Specifically, I argue that although Fox's philosopheme offers a compelling critique of Western consumerism, the apocalyptic dimensions of his critique become deeply problematic when applied to rural Alaska communities. As an alternative, I turn to Seth Kantner's creative nonfiction collection, *Shopping for Porcupine*, as well as Alaska Native narrative traditions surrounding the trickster figure Raven.[8] These traditions encompass a broad spectrum of stories; my argument draws on contemporary representations of Raven (and ravens) by Tlingit and Koyukon Athabascan poets and storytellers.[9] My hope is that such confluences of pre- and postmodernist influences may prompt us to rethink the oft-presumed geographic and temporal peripherality of rural voices to the future of environmental aesthetics, and to search out more inclusive visions of Alaska futurity undefined by tropes of lastness, lostness, or frontier.

Leaving Alaska

Grant Sims's memoir *Leaving Alaska* begins with loss. Feeling alienated from Alaska after its status as prelapsarian wilderness is conceptually eviscerated by the 1989 *Exxon Valdez* oil spill in Prince William Sound, Sims mournfully parts ways with the state, writing: "In the end . . . [we] had found our wildlands to be illusion. The *Exxon Valdez* simply poured oil on them . . . and we were forced to acknowledge that *all that is wild* today is merely indulged . . . merely matters of corporate mercy" (5, my emphasis). Mourning not just a specific place but also the abstract space of romantic "illusion," Sims indicates the scalar logic bound up with environmental elegy, by which deleterious impacts on specific places, such as Prince William Sound, can be imagined to jump scale to encompass the entirety of "all that is wild."

Leaving Alaska is a complicated volume, and valuable not least as a local, political, and environmental history of the oil spill's immediate aftermath. The problem is that in elegizing Alaska as a lost frontier—specifically, a casualty of corporate-legislative cronyism—Sims not only implies, as Morton puts it, that "history is at an end, with the capitalist form as its final destination" but also perpetuates the trope of the Vanishing Indian, which has long worked to delegitimize indigenous futurity.[10] Consequently, the challenge *Leaving Alaska* sets for its successors is to represent the real pain caused by seemingly apocalyptic impacts like *Exxon Valdez* while keeping the future, however compromised or imperfect, more squarely in view.

Sims's disenchantment with the state begins with the death of a Gwich'in boy named Virgil James, whose alcohol-related drowning Sims finds symptomatic of cultural disintegration commensurate with a lost frontier. Sims traces the causes of Virgil's death back 140 years to the establishment of a Hudson's Bay Company outpost that generations later would become his mother Ramona's village. Sims reports that initially "Ramona's people—the Gwich'in, or river people—had no villages," and so their annexation by empire into a society of fixed settlements would leave Ramona, "the child of 15,000 years of Athabaskan roamings along these shores," the "tragic" choice between "the road that led to a future among the white man's shopping malls . . . [or] instead the old Gwich'in road, the river road, to the tiny village that remains today among the most isolated in Alaska" (57–58). Sims offers here his own iteration of the false dilemma we saw Morton identify earlier, between "the idea of a radical outside of the social system" and the "final destination" of late-capitalist modernity. Unable to relinquish her heritage yet also unable to reclaim its nomadic subsistence lifeways, in Sims's view, Ramona James chooses a path without a "future."

In the final chapter, Sims looks back fondly on Alaska from his new home out of state. At this point, *Leaving Alaska* evokes that older elegist, Tennyson, for whom "'tis better to have loved and lost / Than never to have loved at all."[11] Yet working through loss looks very different for the people Sims leaves behind. That difference is acute in the aftermath of Virgil James's death, where

> [t]he frantic and sobering anguish that Eddie James must have felt as he sped time and time again through the creek's frustrating oxbows is only hinted at in the words of a still-drunk Ramona, talking into a tape recorder held by Trooper Roger Maynard. "Soon as we find out Virgil was gone, why Eddie, he just screamed and threw that beer right out of the boat." (65)

Eddie's paroxysm of grief echoes a different part of Tennyson's poem, where in the face of extinction a despondent speaker cries, and cries, "and with no language but a cry" (Canto 54). Futurity for the Jameses and, by extension, for Alaska's indigenous communities, falls outside *Leaving Alaska*'s affective event horizon.

The Cold World

The Ramona James episode is but one of *Leaving Alaska*'s portraits of grief and futility in a lost frontier. Yet we could ask whether tropes of grief, futility, and lostness are as terminal as they may seem. Dominic Fox proposes that in view of the late-capitalist injunction to enjoy, perhaps the opposite experience—dejection, dysphoria—might hold aesthetically and politically revolutionary potential.[12] He terms this state "the cold world," a state of libidinal exhaustion in which the bland promises of modernity—an iPhone in every pocket, a chicken in every pot—are stripped of their enchanting aura:

> The cold world imposes itself as final, terminal, because it is the termination of a world, its metaphorical freezing or blackening. Just as a given "life-world" is endowed with resources and qualities which make it possible to live well within it, so the experience of the cold world (or "unlife world") is the experience of the exhaustion of these resources and the extinction of these qualities. (7)

Fox links this experience of "termination . . . exhaustion" to the figure of "the militant"—"someone who has decisively rejected every source of consolation, every hope that tomorrow will of its own accord come bearing some respite, some unanticipated good news." Relinquishing the possibility of living well, this figure is at last immune to the siren song of capital, and consequently positioned to explore alternative ways of being in the world (43).

His proposition is intriguing. Whereas Tulloch points out elegy's propensity "to inhibit action, paralyzingly wrapping the end into now, heralding the absolute despair of a vague future time," Fox suggests that, paradoxically, a renunciation of futurity—"absolute despair"—may be essential to envisioning alternative futures not already integrated into the compulsorily optimistic late-capitalist imaginary. What Fox does not address, however, is whether the cold world still constitutes a politically transformative experience when we shift its locus northward.

Although the cold world may amount to a provocative mode of resistance to capital's dream of endless markets meeting newly minted desires, realizing that world as Fox describes it would seem to entail significant collateral damage. Because Fox defines the cold world as a totalizing, terminal state, he must extend its borders to encompass "the vitalist world fallen into inert matter, the animist world deserted by its presiding spirits" (6). This invocation of "the vitalist world . . . the animist world" should be familiar. It parallels Sims's scalar leap from Prince William Sound to the entirety of "all that is wild," and it poses similarly problematic consequences for indigenous communities.

Secular, anthropocentric, objectified, Fox's "animist world deserted by its presiding spirits" evokes the ashes of a Western imperial world that has burnt itself out. The problem for northern communities is that this burnout is not just literal, in the form of arctic warming, but also deeply asymmetrical in the pain it inflicts disproportionately on circumpolar species and cultures. The end of the world as Fox frames it may be affectively compelling, but it is a problematic waypoint for communities trying to navigate a changing North.

Shopping for Porcupine

I want to shift now from documenting this problem to considering where literature of the North might propose some potential solutions. Our context will be climate change. Specifically, what would a climate change narrative look like that keeps in view not just the suffering but also the futurity of imperiled communities like those of Grant Sims's Ramona and Eddie James—and, moreover, that contextualizes those experiences not just among humans but also among nonhumans, even the "animist . . . presiding spirits" exiled from Fox's cold world?

To begin with, it would have to be more than anthropocentric. Nancy Lord's recent chronicle *Early Warming: Crisis and Response in the Climate-Changed North* makes clear how unclear environmental ethics become when zoomed out to the scale of climate. Spruce trees are dying but bark beetles are delighting as weather warms, such that what Lord describes as the beetles' "march of death across the landscape" is also a march of life for the grassland species that benefit from habitat succession.[13] This troubling relativity indicates that the pleasure of moral certainty, of definitive solutions, of a blameless way of being in the world, is off the table once we think beyond the human. It provokes writers to address the diversely nonanthropomorphic dimensions of environmental change, not to position them in a false aesthetic unity—that of elegy, or apocalypse, or cautionary tale, or survival narrative—but rather to keep the fantasy of human mastery in check, to recognize that wayfinding might best begin with admitting how lost we are in the midst of the wider-than-human world.

Within recent Alaska writing, that recognition may be most apparent in the work of Seth Kantner. His creative nonfiction collection, *Shopping for Porcupine*, details Kantner's contemporary life in arctic Alaska, caught between Iñupiaq subsistence traditions and late-modern mass culture. The essay "These Happy Spruce" reflects on the fate of Iñupiaq culture imperiled by global climate change. Kantner begins with local accounts of the changing arctic ecosystem:

> all over the region, people are talking of this sun-drenched summer . . . talking of the strange occurrences of a walrus swimming upriver . . . harbor seals as far into fresh water to Onion Portage, and beluga coming at the wrong time, in midsummer, to Kotzebue sound. (197)

He adds other examples, such as open tundra lost to brush and spruce trees through habitat succession, and then shifts into a detailed memorial for two elders who vanished while seal hunting on increasingly thinning ice.

Here the essay rises to an elegiac crescendo. "Now the names are piling up," he writes, of residents "drowned through weak ice, thin ice, refrozen ice . . . And we have become familiar with that ominous black cloud on the horizon—the reflection of open ocean on winter sky" (207). Alluding to that other black cloud of atmospheric carbon, Kantner offers a compendium of loss. We seem to be back in Grant Sims's territory, elegizing Alaska as a lost frontier. Yet just as the "These Happy Spruce" seems poised to move toward resolution, the essay plunges through its own thin ice of narrative reliability. The last lines read:

> Ironically, warmth, on the surface at least, is not hard to accept. Especially in the Arctic. With it, I'm afraid, we may have to accept many other things: funerals not just for people but also for whole species and foods and familiar ways we love—such as traveling on ice. Yet a hungry land forces acceptance of what the land serves up. From my dad's stories . . . I know that he admired the old Eskimos' optimism in facing an uncertain tomorrow, and before dusk I walk back down the trail to his log cache, to eat a few more sweet irresistible berries before they freeze, and to look again at those happy trees spreading across the tundra to the north. (207)

As usual in Kantner's writing, this passage serves up a superficial kind of resolution only to undermine it immediately, foregrounding the writer and resident's struggle to make sense of his changing homeland. On the one hand, it seems that the land will reliably provide, "serv[ing] up . . . sweet irresistible berries" and "happy trees." Yet this same gesture of optimism and acceptance is framed as "ironic" at the outset, just as the elders' optimism is framed as well within "my dad's stories." Outside these comfortable narrative frames, sometimes "a hungry land" serves up nothing, bound to its harsh cyclicity of arctic booms and busts.

For this reason, even though Kantner seems to fall back into pastoral complacency at the essay's end, "These Happy Spruce" is by no means a back-to-nature narrative. Not when a bumper crop of blueberries thriving in the warming weather and a black cloud of carbon emissions are literally the same thing, both fractal nibs on the nonedge of what Timothy Morton has called the "hyperobject" of global warming.[14] The irreconcilable heterogeneity that Kantner's Arctic embodies is difficult to accommodate within any kind of narrative closure, and so "These Happy Spruce" downshifts its conclusion to the level of literary form. As much as Kantner resists elegizing Alaska as a lost frontier, his passage also critiques the opposite affective pole of "sweet . . . happy" optimism by stretching it to the breaking point, to the point where a few blueberries and interspecies empathy are being asked to stand up, ultimately, to extinction.

The ironic dissonance with which the essay ends breaks the spell of authorial authority, laying bare the text as a difficult negotiation between audience desire and Kantner's lived experience. It calls into question readerly expectations of affirmation, instruction, and ultimately coherence in the face of climate change. Kantner can neither mourn the loss of tundra nor can he uncritically accept "these happy spruce" advancing North. As an artist, he can only emphasize the irony. Consequently, "These Happy Spruce" is as much about writing about climate change as it is about climate change itself. In a letter, Kantner describes himself as contorted by impossible, conflicting demands of speaking for a culturally contested, ecologically uncertain terrain within additional demands of audience and market. He explains, "in the end probably all those forces make me rewrite every word 400 times instead of 4. (I'm not exaggerating.) Maybe the forces are good. They are not fun, though."[15] Kantner can't go on, he'll go on. Here, in that self-reflexive, conflicted doubleness, not the finality of leaving, or the clarity of cold world dejection, lies the really difficult work of Alaska elegy.

Ravens' World

"These Happy Spruce" exemplifies a shift in recent Alaska writing away from terminal essentialisms of last wilderness and lost frontiers to a perspective that is far more cosmopolitan, more contested, and more contemporary than we often associate with literature of the North. Yet something about this argument might strike us as troublingly teleological, in its implication that ambivalent, ironic texts like Kantner's indicate that literature of the North has at last left behind its local color beginnings to "come of age" or "catch up" with literary theory. I want to end by asking if there is another way to understand Kantner's position in relation to literary modernism and modernity.

This entails reading beyond the Western canon. We could start among Alaska Native traditions surrounding Raven, an outrageous trickster who revels in impurity and irony. Storytellers make clear Raven's debauched nature—he cheats, steals, and delights in manipulating his friends. Yet Raven also created humans, brought light to the world, and performed other originary miracles.[16] In his novel *Raven Stole the Moon*, Garth Stein frames Raven's ambiguous nature this way, through a conversation between a modern-day Tlingit shaman and a consternated white protagonist:

> "Do you understand, Ferguson? Raven didn't just give us the sun, moon, and stars. He had to steal them from someone else."
>
> "I don't follow."
>
> "Stealing is an act of evil. But giving is an act of good. So was Raven good or evil?"
>
> Ferguson felt a little dumb for having to be led to the answer.
>
> "Both."[17]

As an ecological being and a cross-cultural discursive construction, a figure both sacred and scandalous, a species ancient and contemporary, at home across hemispheres, Raven confuses the boundaries of local and global, natural and anthropogenic, mythic and historical, base and sublime.

I cannot help, therefore, but compare Raven to another mythic figure, Donna Haraway's posthuman cyborg, who takes "pleasure in the confusion of boundaries and responsibility in their construction."[18] Haraway offers this "ironic political myth" to document the ways in which contemporary humans are already cyborgs, late-modern hybrids of human and animal, organism and machine, material and information (149). Yet as Haraway writes that "cyborg politics insist on noise and advocate pollution, rejoicing in the illegitimate fusions of animal and machine," one can also hear the raucous chatter of real ravens jostling on an urban lamppost, or rifling through a rural dump, or rejoicing over a roadside moose kill (176). Like the cyborg, or the postmetaphoric polar bears of Allison K. Athens's contribution to this volume, Raven/ravens "skips the step of original unity" (151), renouncing the notion that, as Athens puts it, "there was once an originary moment that was more authentic than the present one." Raven/ravens exist from before the beginning as Stein's conflicted figure, thief and giver "both."[19]

I want to suggest what these parallels between pre- and postmodernist perspectives might mean for northern studies, and especially for Western humanists interested in a comparativist literary North. Yet if we are going to position Raven/ravens as the North's own "ironic political myth," I have to acknowledge first the prickly consequences of making that claim. In an interview, Nancy Lord addresses Alaska's "not-very-happy history of outsiders (writers, anthropologists, scientists) coming into villages to 'take' stories and expertise, often in a way that seems exploitive to the local people."[20] Given this "not-very-happy history," the challenge for our discussion is to think Raven/ravens and cyborgs together without arrogating the former to the later, without railroading Alaska Native traditions into one big intellectual free-trade zone.

It's a challenge perhaps Western humanists can measure up to, not by confidently appropriating Raven into a closed critical discourse but rather by following real ravens' lead. The inquisitiveness often associated with *Corvus corax*'s species survival is tempered by an extreme wariness that might translate for our purposes into both audacity and humility.[21] Striking that balance in scholarship means lingering in the generative uncertainties and owning up to the inevitable faux pas and hypocrisies, which unfold with every step into the challenging territory of transculturation. It also means going beyond the scope of this essay to engage directly with the work of contemporary Alaska Native storytellers and artists—without which this entire proposition amounts only to more well-intentioned cultural appropriation.

Hence the plurality of Ravens in this essay's title. Introducing their collection *Environmental Criticism for the Twenty-First Century*, the editors assert, "If it is not possible for ecocritics to reject metaphor altogether in their interdisciplinary projects . . . then at least we might choose metaphors . . . that are lively agents, tropes which turn away from fantasies of pure referentiality."[22] I approach Raven/ravens here cautiously, as a metaphor that conceptually holds together just enough to emphasize its internal disjunctions. As a zoological object, a charismatic megafauna, a sacred being, an ecocritical metaphor, a cypher, a pest, Raven/ravens are poised to put each of those identities' constituencies together into dialogue—the point at which this exercise becomes more than academic.

Bringing her "Cyborg Manifesto" to a close, Donna Haraway avers, "Though both are bound in the spiral dance, I would rather be a cyborg than a goddess" (181). It is ironic that a postmodernist manifesto militating against binary constructions of all stripes would end with this pithy dualism, but I'd like to borrow it nevertheless to pose a similar choice for northern literary studies. In the face of world-upending phenomena like global climate change, this would be the choice between (on the one hand) elegizing Alaska as a lost frontier and (on the other) reaching out as widely as possible, transculturally, interdisciplinarily, to recompose Alaskan futurity. Though both are bound in visions of the North, I would rather be a raven than a ghost.

Acknowledgments

Thanks to the editors, Mike Ziser, Tim Morton, and Hsuan Hsu.

Endnotes

1. Grant Sims, *Leaving Alaska* (New York: Atlantic Monthly, 1994). Future citations appear in text. James Campbell, *The Final Frontiersman* (New York: Penguin, 2002); Jennifer Brice, *The Last Settlers* (Pittsburgh: Duquense University Press, 1998); Sheila Nickerson, *Disappearance, a Map: A Meditation on Death and Loss in the High Latitudes* (New York: Doubleday, 1996).

2. Eric Heyne, "From 'the Last Frontier' to *The Island Within*: Two Versions of Alaska in Contemporary Nonfiction Narrative," *The Northern Review* 21 (Summer 2000): 38–56.

3. Eric Heyne, "'Such Humble Awareness': The Emergence of a Northern Vision in the Fiction and Non-Fiction of Nancy Lord," *The Northern Review* 27 (Fall 2007): 66.

4. William Hunt, "Between Myth and Apocalypse," *The Nation* 7 (January 1978): 22. Future citations appear in text.

5. Timothy Morton, "Don't Just Do Something, Sit There! Global Warming and Ideology," in *Rethink: Contemporary Art and Climate Change*, ed. Anne Sophie Witzke (Copenhagen: Alexandra Institute, 2009), 49–52.

6. Susan Kollin asserts Alaska's exceptional status in *Nature's State: Imagining Alaska as the Last Frontier* (Chapel Hill: University of North Carolina Press, 2001).

7. Erica Silverstein, "8 Compelling Reasons to Cruise to Alaska Now," in The Independent Traveler, Inc., *Cruise Critic,* 2011, http://www.cruisecritic.com/articles.cfm?ID=1237.

8. Seth Kantner, *Shopping for Porcupine: A Life in Arctic Alaska* (Minneapolis: Milkweed Editions, 2008). Future citations appear in text.

9. Specifically, I refer to Tlingit perspectives from Robert Davis's poetry collection *Soulcatcher* (Juneau: Raven's Bones Press, 1986) and Nora Dauenhauer's short plays in *Life Woven with Song* (Tucson: University of Arizona Press, 2000). Richard Nelson's *Make Prayers to the Raven: A Koyukon View of the Northern Forest* (Chicago: University of Chicago Press, 1983) presents Nelson's ethnographic interpretation of beliefs concerning Raven; Koyukon elder Catherine Attla's collection *Bekk'aatugh Ts'uhuney: Stories We Live By* (Fairbanks: Yukon Koyukuk School District and Alaska Native Language Center, 1996) similarly represents Raven as a powerful, paradoxical figure.

10. Brewton Berry, "The Myth of the Vanishing Indian," *Phylon* 21, no. 1 (1960): 51–57.

11. Alfred, Lord Tennyson, "In Memoriam A. H. H. OBIIT MDCCCXXXIII," Canto 27. Future citation appears in text.

12. Dominic Fox, *Cold World: The Aesthetics of Dejection and the Politics of Militant Dysphoria* (Winchester, UK: Zero Books, 2009). Future citations appear in text.

13. Nancy Lord, *Early Warming: Crisis and Response in the Climate-Changed North* (Berkeley: Counterpoint, 2011), 25.

14. Timothy Morton, *Hyperobjects: Philosophy and Ecology after the End of the World* (Minneapolis: University of Minnesota Press, 2013).

15. Seth Kantner, message to the author, February 2012.

16. In Joseph Bruchac, ed., *Raven Tells Stories: An Anthology of Alaska Native Writing* (Greenfield Center, NY: The Greenfield Review Press, 1991).

17. Garth Stein, *Raven Stole the Moon* (New York: HarperCollins, 1998), 46.

18. Donna Haraway, "A Cyborg Manifesto: Science, Technology, and Socialist-Feminism in the Late Twentieth Century," in *Simians, Cyborgs and Women: The Reinvention of Nature* (New York: Routledge, 1991), 150. Future citations appear in text.

19. On the potential problems with posthumanist enthusiasm for renouncing origins and authenticities, see for example Zakiyyah Iman Jackson, "Animal: New Directions in the Theorization of Race and Posthumanism," *Feminist Studies* 39, no. 3. (2013), and "Outer Worlds: The Persistence of Race in Movement 'Beyond the Human,'" *GLQ: A Journal of Lesbian and Gay Studies* 21, no. 2–3 (2015): 215–218.

20. Holly Hughes, "Finding Beauty in a Troubled World," *Terrain: A Journal of the Built Environment* 30 (Fall 2012), http://terrain.org/2012/interview/interview-with-nancy-lord/.

21. Lawrence Kilham, "Roadkills and Fear Reactions" and "Reaction to Snow," in *The American Crow and the Common Raven* (College Station: Texas A&M University Press, 1990).

22. Stephanie LeMenager, Teresa Shewry, and Ken Hiltner, *Environmental Criticism for the Twenty-First Century* (New York: Routledge, 2011), 3.

Bibliography

Alfred, Lord Tennyson. "In Memoriam A. H. H. OBIIT MDCCCXXXIII." Canto 27.

Attla, Catherine. *Bekk'aatugh Ts'uhuney: Stories We Live By*. Rev. ed. Transcribed by Eliza Jones. Translated by Eliza Jones and Chad Thompson. Fairbanks: Yukon Koyukuk School District and Alaska Native Language Center, 1996.

Berry, Brewton. "The Myth of the Vanishing Indian." *Phylon* 21, no. 1 (1960): 51–57.

Brice, Jennifer. *The Last Settlers*. Pittsburgh: Duquense University Press, 1998.

Bruchac, Joseph, ed. *Raven Tells Stories: An Anthology of Alaska Native Writing*. Greenfield Center, NY: The Greenfield Review Press, 1991.

Campbell, James. *The Final Frontiersman*. New York: Penguin, 2002.

Davis, Robert. *Soulcatcher*. Juneau: Raven's Bones Press, 1986.

Dauenhauer, Nora. *Life Woven with Song*. Tucson: University of Arizona Press, 2000.

Fox, Dominic. *Cold World: The Aesthetics of Dejection and the Politics of Militant Dysphoria*. Winchester, UK: Zero Books, 2009.

Haraway, Donna. "A Cyborg Manifesto: Science, Technology, and Socialist-Feminism in the Late Twentieth Century." In *Simians, Cyborgs and Women: The Reinvention of Nature*, 149–181. New York: Routledge, 1991.

Heyne, Eric. "From 'the Last Frontier' to *The Island Within*: Two Versions of Alaska in Contemporary Nonfiction Narrative." *The Northern Review* 21 (Summer 2000): 38–56.

———. "'Such Humble Awareness": The Emergence of a Northern Vision in the Fiction and Non-Fiction of Nancy Lord." *The Northern Review* 27 (Fall 2007): 60–76.

Hughes, Holly. "Finding Beauty in a Troubled World." *Terrain: A Journal of the Built Environment* 30 (Fall 2012), http://terrain.org/2012/interview/interview-with-nancy-lord/.

Hunt, William. "Between Myth and Apocalypse." *The Nation*, January 1978.

Jackson, Zakiyyah Iman. "Animal: New Directions in the Theorization of Race and Posthumanism." *Feminist Studies* 39, no. 3 (2013): 669–685.

———. "Outer Worlds: The Persistence of Race in Movement 'Beyond the Human.'" *GLQ: A Journal of Lesbian and Gay Studies* 21, no. 2–3 (2015): 215–218.

Kantner, Seth. Message to the author, February 2012.

———. *Shopping for Porcupine: A Life in Arctic Alaska.* Minneapolis: Milkweed Editions, 2008.

Kilham, Lawrence. "Roadkills and Fear Reactions" and "Reaction to Snow." *The American Crow and the Common Raven.* College Station: Texas A&M University Press, 1990.

Kollin, Susan. *Nature's State: Imagining Alaska as the Last Frontier.* Chapel Hill: University of North Carolina Press, 2001.

Morton, Timothy. "Don't Just Do Something, Sit There! Global Warming and Ideology." In *Rethink: Contemporary Art and Climate Change*, edited by Anne Sophie Witzke. Copenhagen: Alexandra Institute, 2009.

———. *Hyperobjects: Philosophy and Ecology after the End of the World.* Minneapolis: University of Minnesota Press, 2013.

———. *The Ecological Thought.* Cambridge, MA: Harvard University Press, 2010.

Nelson, Richard. *Make Prayers to the Raven: A Koyukon View of the Northern Forest.* Chicago: University of Chicago Press, 1983.

Nickerson, Sheila. *Disappearance, a Map: A Meditation on Death and Loss in the High Latitudes.* New York: Doubleday, 1996.

Silverstein, Erica. "8 Compelling Reasons to Cruise to Alaska Now." *Cruise Critic.* The Independent Traveler, Inc., 2011. http://www.cruisecritic.com/articles.cfm?ID=1237.

Sims, Grant. *Leaving Alaska.* New York: Atlantic Monthly, 1994.

Stein, Garth. *Raven Stole the Moon.* New York: HarperCollins, 1998.

3

The "Bear Who Began It" and the Metaphorics of Climate Change

allison k. athens /
UNIVERSITY OF CALIFORNIA, SANTA CRUZ

It was the bear who began it. Said,
I'm getting out from under.
I am not Bear, l'Ours, Ursus, Bäṛ
or any other syllables
you've pinned on me.
[. . .]
I renounce metaphor: I am not
child-stealer, shape-changer,
old garbage-eater, and you can stuff
simile also: unpeeled,
I am not *like a man.*

I take back what you have stolen,
and in your languages I announce
I am now nameless.
My true name is a growl.[1]
—*Margaret Atwood*

RENOUNCING METAPHOR, THE BEAR IS THE FIRST TO REJECT HER name in Margaret Atwood's poem "The Animals Reject Their Names and Things Return to Their Origins." However, even as the bear goes nameless, she cannot escape announcing her new onomatopoeic name—growl—in language and in the poem, both of which rely on metaphor to communicate and connect across semantic, cognitive, ontological, and bodily difference. Through words and phrases that signal otherness, the language of the poem separates and unites disparate ontologies. The bear who begins it is a testament to a connection that cannot escape the binds of meaning even as she returns to a moment before categorization had the potential for perilous consequences. Although the animals in Atwood's poem are on a quest for an originary moment before language bound them to the roles and limitations of acting as and for humans, the transformative processes they undergo reveal that change is never the same as erasure.

Change, transformation, adaptation, and disappearance: these are the terms of species and environments caught in an era that some scholars are calling the Anthropocene, an epoch in which the effects of human activities are etched in the geologic record.[2] Bearing a striking resemblance to Fred Bodsworth's Eskimo curlew presented in Elspeth Tulloch's careful analysis of "ecoelegy" at the beginning of this section, David Attenborough follows another increasingly fragile species who metonymically embodies the geologic shift toward extinction by human activity: the polar bear. In his *Planet Earth* television series, the film crew catches a rare sight of a polar bear swimming in pack ice. As the helicopter retreats and the camera zooms out, Attenborough tells the viewer that she is witnessing the spring breakup of the ice pack, when male polar bears swim out to sea in search of food while female bears stay on land with their cubs. Although the connection to a warming world is not made in his speech, the implication is there through the feeling of loss that permeates the cinematography. The encounter among film crew, Attenborough, audience, and bear is a meeting to say good-bye.[3] Polar bears, in my story and that of *Planet Earth*, are increasingly out of place and, in the elegiac narrative of the Anthropocene, potentially out of time. However, unlike Attenborough's bear, the polar bear I discuss in the following essay is not a bear who took to the seas but one who wandered away from her home in the northern coastal Arctic to the inland southern (or sub-) arctic community of Fort Yukon, Alaska, where conventional and traditional wisdom about polar bears agrees she does not belong. The bear who wandered down from the North was a disturbing presence in the subarctic territory of Fort Yukon's Gwich'in Athabascans, who are more used to encounters with grizzly bears. Furthermore, her arrival had the unsettling effect of calling attention to the psychogeographic parameters of the Gwich'in homeland, affective borders that are constantly being negotiated in contemporary political and environmental administrative practices. For Attenborough, the Gwich'in, and many people who view the polar bears' struggle to adjust to a warming planet as indicative of human failure to care for the world, bears out of place reveal the crisis at hand: human-induced climate change has just about made the planet unlivable.

Relying on the fixed categorization of species and environments, stories told in the mode of crisis threaten that a complete ecologic loss is imminent and it is almost, or is, too late to rectify the situation. And if it is too late, it hardly seems worthwhile to engage in the difficult process of ecologically minded change. In contrast to the paralysis of inevitability, an attention to microstories and alternative storytelling paradigms suggests opportunities for continued relationships with polar bears and northern ecologies, even if, in a not-too-distant future, they exist in an altered form. By acknowledging and being open to transformation, the messy microconnected relationships of the here and now may offer other ways of inhabiting and caring for the world. Or, in Will Elliott's evocative phrasing in the previous essay, "reaching out as widely as possible, transculturally, interdisciplinarily, to recompose Alaska futurity" will allow us to become ravens instead of ghosts.

Rather than telling the bear's story of change in the apocalyptic mode that treats her as an allegory of impending catastrophe and loss, what happens when the story is told in the form of a travel narrative? This would be a story not about the finality of the bear's death (and she does die in the home of the Gwich'in) but about migration, a story of movement, return, and transformation. Change in this mode of narration is more than the destruction of all that is known and protected, because the outcomes of change cannot be known in advance. In one environment (the northern arctic coast of Alaska), the polar bear is a white bear that inhabits a marine ecosystem, a predatory mammal dependent on the ocean for her prey (seals) and the snow and ice for her hunting techniques and denning sites. When she travels south, her definition is a little less clear. The bear's physical trek across the formidable terrain of Alaska is also a trek across domains of signification. Following the polar bear across narrative modes and material landscapes situates the Arctic and subarctic as spaces that are variegated, relational, contingent, and locally produced and meaningful.

Focusing on the science of snow and ice and the repercussions of melting temperatures for the Iñupiaq inhabitants of northern Alaska, Charles Wohlforth writes in detail about climate change for the liminal zone of arctic land and ice edges. In passing, he also discusses the subarctic area of the Arctic National Wildlife Refuge (ANWR) and the effects of drilling pads and pipelines on caribou herds.[4] Tellingly for the subarctic, Wohlforth reports after a conversation with an environmental scientist working at the University of Alaska's Toolik station in the foothills of the Brooks Range: "ANWR was just more wet tundra and mosquitoes, the same swampy terrain that stretched for hundreds of miles east, west, and north of us . . . [the graduate student scientist had] expected special nature to be more scenic."[5] Highlighting the environmental movement's strategy for invoking the uniqueness of a region as a reason for its preservation, the graduate student is—when actually within the landscape—unsure what to feel when her expectations are unmet. The graduate student's experience questions whether aesthetic language can ever adequately be used to engage with the world, whether sentiment can be the basis for political and ethical action. For Wohlforth, the "landscape's story" of the subarctic only gains "a beginning, middle, and end" when the Dalton Highway and the Trans-Alaska pipeline add a "unifying theme."[6] In this narration, the swampy subarctic of the Arctic National Wildlife Refuge remains an illegible wilderness outside an ethics of care given that on its own it lacks a narrative structure.

The difference between the treatment of the Arctic and the subarctic in critical discussions is not, I argue, on account of privileging the most destructive effects of a warming planet, such as the loss of arctic ice and polar bears. Rather, the figural language that has produced "North" in the literary imagination has not had room for the muddier and less iconic subarctic. Along with areas in the Arctic, the subarctic is also melting and the region's flora, fauna, and culture are under threat of irreversible change. Nonetheless, Wolforth and the graduate student scientist cannot find a frame of reference from which to make sense of what is being viewed or experienced in connection with the subarctic landscape. While the scientist is an expert in her field of microcosmic botanical identification and tundra plot analysis, the larger scene is both overwhelming and underwhelming: it cannot be narrated. Addressing this challenge through the framing devices of his photographs, photographer and climate advocate Subhankar Banerjee offers a way in which to come to terms with vast space. He simultaneously acknowledges and refutes the standard tropological framework governing how nonarctic people perceive arctic and subarctic landscapes. He reframes the two regions through the lens of inhabitation, rather than through discourses of awe or emptiness, sublime or waste area, pure or polluted. His photographs of people and animals engaged in the labor of successful inhabitation (herders surrounded by their reindeer, the earth permanently seamed from the tracks of thousands of migrating caribou, a mother polar bear with cubs) brings the two climatic regions together through how people and animals mutually live in these northern spaces.

Banerjee's photographs and testimonial work in the subarctic documenting Gwich'in lives in and around the Arctic National Wildlife Refuge reveal a place as contested and transformative as the north coastal plain. He offers a new frame for imagining North beyond the Romantic sublime, exemplified in his inclusive approach of "land as home" and "ecocultural rights."[7] Ecocultural rights encompass an understanding of lived environments and include a human right to a flourishing life and the right to exist for other species, systems of land, seas, and watersheds.[8] In short, it is an ecosystemic (or in Elliott's terms "ecosocial") approach that takes into consideration the fact that humans—and human culture—shape the understanding and ability to live on and with the land but that land is, fundamentally, necessary for human life and culture. Banerjee's work in the subarctic demonstrates care for the world while it undergoes dramatic transformation, and it opens up possibilities for reading the story of the Fort Yukon polar bear as something other than the tragic loss of a climate refugee.

In the early spring of 2008, an incident occurred outside the subarctic village of Fort Yukon, a Gwich'in Athabascan community located on the confluence of the Porcupine and Yukon Rivers in the Interior of Alaska. Situated more than two hundred and fifty miles inland from the arctic coast and between two mountain ranges, the Brooks Range to the north and the Alaska Range to the south, the village became the epicenter for a dramatic conflict involving history, the social imagination, colonialism, the petroleum industry, and environmental activists. A young female polar bear wandered across the arctic plain and over the northern mountains and was shot and killed by a Gwich'in hunter along a trapline near the community.[9] Crossing the Brooks Range is no small accomplishment, given that the mountains not only divide the landscape but also function as a physical, psychological, and cultural barrier between the Iñupiaq Eskimos who live along the arctic coast and the Athabascan Indians who inhabit the subarctic tundra.[10] The historical tensions that exist between the Gwich'in (Athabascan Indians) and Kaktovikmuit (Iñupiaq Eskimos), recently exacerbated by a colonial governing system and disagreements about oil drilling in the Arctic National Wildlife Refuge, are pertinent to the discussion of this polar bear.[11] Each group, furthermore, seeks a modernity that does not rely on the administrative practices of a government that resides elsewhere and consistently fails to understand Iñupiat and Athabascans as historical *and* modern people.[12] This wayward polar bear is more than simply a being in the wrong place at the wrong time. In her very physicality, the bear also becomes the figure upon which the effects of climate change and a fight for political legibility turn.

In her study of the "social lives of stories," Julie Cruikshank observes that "[n]arratives provide a range of viewpoints from which to interpret the discrepant meanings we encounter both in human history and in daily experience. Through narrative plots, both chronological sequences and recognizable patterns are revealed."[13] One textual framework that provides a reference point for the Gwich'in when involved with bears acting oddly is the story of the "ice bear." In Moses Peter's version of "The Ice Bear," the bear, or "*ch'atthan* as the Kutchin [Gwich'in] Indians call it" demonstrates foreknowledge of both time and environment. The ice bear, Peter relates, is "a grizzly bear that doesn't hibernate in winter like other bears. When he doesn't hibernate, he knows he's in danger, so he goes down to some open water and dips himself in. He dips himself in again and again to make a thick armor of ice all over his body. It is hard to kill a bear like that because of all the ice on his fur."[14] The ice bear story reveals how one might fashion an understanding between humans and nonhumans that refrains from sentimentalizing the animal-other as an always-known-in-advance entity.

Barry Lopez also attempts to account for animals acting in unexpected ways, but from the position of a Western scientist-observer, a perspective that comes up lacking: "We are sometimes at a loss in trying to describe such [unaccountable reactions in predator-prey relations] because we unthinkingly imagine the animals as instinctual. We are suspicious of motive and invention among them."[15] The grizzly/ice bear in the story must react to an unperceived event, the inability to hibernate. The young hunter must also adapt to a novel situation while he decides, in the midst of his encounter with "the thing" drawing ever nearer, what is worth saving for some sort of future. In an effort to return to his village to warn of the ice bear, the hunter first sacrifices his friend to the bear, then his material possessions (arrows, shirt, ax), and, finally, on reaching his home, himself. He cries out, "'There's an ice bear! A *ch'atthan*!' That was all he had a chance to say, for just then blood started pouring from his mouth. He must have frozen his lungs from running. He died right there."[16] The bear is a thinking, rational being with intentionality who takes action in the world, and when he does, both bear and humans are changed in the encounter. Although both die, the relationship between bears and Gwich'in lives on in the story. "The Ice Bear" provides Gwich'in Athabascans a structure for accommodating and interpreting a novel situation like the polar bear near Fort Yukon, even if the narrative cannot fully mediate the arrival of, or even make acceptable, all forms of change.

A common Gwich'in storytelling invocation describes human and nonhuman communication: "In the ancient days, when all of the people could talk to the animals, and all of the animals could speak with the people." The Gwich'in hero, Vasaagihdzak, is famous for profiting from this ability to communicate with other animals and for ending up the butt of many jokes played on him by these same creatures. In her written version of the oral tale, Katherine Peter explains that Vasaagihdzak was "famous for helping people. He also helped animals. He could speak to each animal in its own language. He could even speak to the trees. This is how he helped the people."[17] The stories about Vasaagihdzak's connection to the animals illustrate how symbolic narrative systems transmit and produce relationships in real-life encounters between Gwich'in and their relational others: animals, trees, land. With the arrival of the bear in Fort Yukon, however, communication and reciprocity were brought into question. The day after the arrival of the polar bear, Trimble Gilbert—the traditional chief and pastor of Arctic Village, a Gwich'in Athabascan settlement just north of Fort Yukon in the foothills of the Brooks Range—told his congregation: "I hear there is a polar bear in Fort Yukon. Things are changing, you can see. All of you have to be careful. Don't let your children go out by themselves. The animals are no longer the same. They are getting wild. They behave strangely. Watch out for the bears. They don't listen to us anymore."[18] Robert Brightman suggests that bears and bear stories indicate a profound ambivalence about the boundaries of human and animal, because the "signs of the covert infra human animal" are most visible in the bear.[19] A bear that changes her behavior and starts seeking new modes of living in the world can have a profoundly unsettling effect on those around her, especially when lines of communication appear to be broken.

Rather than zooming out from the bear like the BBC camera crew invoked at the beginning of this chapter, the writers of Northern Arizona University's Institute for Tribal Environmental Professionals resist the mode of elegy as they focus on new forms of "northern futurity" exhibited by northern bears. "Polar bears, being bad terrestrial hunters," they explain, "will need to find a new way to exist as the amount of sea ice decreases. Human beings may have to adapt along with the bears and realize this [kind of encounter] may become a frequent occurrence."[20] The polar bear in this story is not seen as dying out and in need of saving by the intervening powers of, for example, Coca-Cola's campaign called "Arctic Home."[21] Rather, the authors underscore the bear's agency as she changes her habits to fit a novel situation. Humans, like this bear, are not exempt from the need (in Lopez's terms) to "test the landscape" and adapt to changes in the world. Some forms of connection and ways of living may come to an end, but this is hardly a death sentence. Rather, new forms of existing—together—in the world can be found. As the NAU piece concludes:

> According to natural history, polar bears originally descended from brown bears. This species started moving north and so began to evolve a different shape of snout, hair color, as well as different hunting strategies in order to adapt to its new habitat. Perhaps one day this species will come again to where it began, completing its circular life history.[22]

The passage does not put the bear in the realm of “nature” while reserving “culture” for the human species also implicated in the event of the encounter in Fort Yukon. Both species are part of a web of movement that sees adaptation not as a biological imperative (evolve or die; only the fittest survive) but as a learning process, passed one to another in a culture of exchange. Perhaps, even, the rediscovery of some old forms of technology and art (such as storytelling) would allow a “completion of a circular life history” for the bear and the Gwich’in, who are caught within colonial administrative systems that legislate how they must inhabit their homeland.

While climate change is often described as a linear movement toward extinction for some species, loss of habitat for others, and the eradication of traditional forms of living, this is only the temporality of a single human life sometimes extended into the future through an appeal to the Earth that children will inherit. Considering the circularity of life histories offers more for connections in the present than discourses of inevitable extinction in the time of the Anthropocene, narratives of crisis in the mode of the jeremiad, or stories of irretrievable and melancholic loss. Complementing Banerjee’s work for ecocultural rights, Donna Haraway proposes that we “stay with the trouble” in the present of lived and ongoing histories of colonization and processes of adaptation.[23] How we read the story of northern presents affects how we enact northern futurities; a different storytelling mode is what Tulloch suggests is needed in “Whose Arctic? Who Cares?”: a “satiric rendition of [the] continuing story,” she writes, might “invite other ways to envision [the story], other points of emphasis—not on endpoints, not on new beginnings—but on the absurd reality bare facts reveal.” Remaining attentive to the "noncontemporaneous time" in the stories we tell about the North might just help guide our understanding of human and animal relations in a given location for uncertain, but possible, futures.[24]

The story of rapid change is not just an arctic tale; it is also deeply felt in the subarctic through the wandering of our hero, the polar bear. In my telling of the story of climate change in Alaska, the Arctic and the subarctic are brought together to more fully grasp the implications of using the polar bear as the icon of climate change. While many scientists are documenting the movement of species north, the polar bear is heading in the opposite direction. In 2007, the United States Geological Survey published a press release about a long-term study in which it had documented a high rate of change in denning sites for female polar bears from a preference for ice to that of land.[25] The report cites reasons for this change: "Arctic pack ice has formed progressively later, melted earlier and lost much of its older and thicker multi-year component. Together, these changes have resulted in pack ice that is a less stable platform on which to give birth and raise new cubs."[26] For the twenty years of this study, which includes several generations, female bears have been testing the world for new places to live. While ice is an integral component of what a "polar bear" is believed to be, the term *polar bear* is actually a metaphor. "Polar bear" describes the being that within a given time and spatial arrangement exhibits certain characteristics, such as preying on seals, living on ice, and being colored white.[27] The polar bear, endemic and iconic mammal of the Arctic, is, as Lopez has written, "a creature of arctic edges: [s]he hunts the ice margins, the surface of the water, and the continental shore."[28] But what if "polar bear" renounces these trappings? The polar bear is an uneasy bearer of human reaction to climate change. While the whiteness of the bear may slowly change to brown, it may be precipitous to begin to eulogize the polar bear. Disconnecting responses to climate change from the tragedy of losing "polar bear" opens the possibility of microconnections to local changes and the stories that relationally do not fit into conventional wisdom but are experienced nonetheless.[29]

Renouncing metaphor, Atwood's bear rejects the fixed categories of species, temporality, and experience to reveal how species, environments, and narratives are always-already connected through shared practices of living. In this respect, language, specifically the figural language of storytelling and poetry, can be used as a tool for connection rather than division. Thinking of Banerjee's "ecocultural rights" as stemming from his work to bring "arctic voices" into scientific and policy-making discussions about climate change reveals that storytelling is an integral mechanism for relational equality.[30] Ecocultural rights encompass more than the seemingly similar iteration of human rights that focus on fixed notions of subjectivity and access to dominant economic and political modes of legibility. Instead, ecocultural rights can bear transformation and adaptation as long as the mode of discourse accounts for the liveliness of all participants—bears, people, caribou, sedges, even botflies—in the humpy, swampy, scrubby, buggy, and lived-in home of the subarctic.

Lopez believes the polar bear to be a "creature of arctic edges." What does it mean to live on the edge? The world's edge is Ultima Thule, the no-man's-land between the real and the fictive, the known and the not known, history and mythology.[31] Climate change is not only the loss of the metaphor of North; it is also the uncanny arrival of a discontinuity with place that disrupts thinking of "land as home." The narratives of climate change provide a vehicle for rethinking relationships to the world and ask that we reconsider what it means to inhabit a location. Attenborough, climate scientists, and ecologists all care for the world even as they see the end of the world in sight. This response is insufficient for a full consideration of ecocultural rights: while adaptation is difficult and the outcome uncertain, to narrate changing lives as lives lost only reinforces that there was once an originary moment that was more authentic than the present one. This is a familiar discourse that functions all too well in the temporality of colonial erasure that stipulates any deviation from a static cultural or biological state is grounds for the removal of all consideration and protection. Humans and bears will have to learn to adapt to new ways of being—and thinking—in the world. A more ecocultural extension of rights to animated beings previously denied consideration (bears, ice, tundra, to name a few) will perhaps, one day, bring all species to where they began when they had languages that spoke across difference, in a completion of circular life histories.

Endnotes

1. Margaret Atwood, "The Animals Reject Their Names and Things Return to Their Origins," in *The Tent* (New York: Doubleday, 2006), 78–79.

2. Paul J. Crutzen and Eugene F. Stoermer, "The Anthropocene," *IGBP Newsletter* 41 (2000): 17.

3. "Ice Worlds," in *Planet Earth: As You've Never Seen It Before: The Complete Series*, produced by Alastair Fothergill (2005; London: BBC, 2006), DVD.

4. The subarctic is a climatic region generally situated between 50°N and 70°N latitude. It is characterized by boreal and black spruce forests, muskeg, permafrost, and animal species such as beaver, moose, grizzly bears, and tree squirrels. However, "sub" has its own connotations; given that the subarctic lies below the arctic zone, suggestions of "lesser," "unimportant," "substandard," and "deficient" suffuse discourses involving comparisons of the two regions.

5. Charles Wohlforth, *The Whale and the Supercomputer: On the Northern Front of Climate Change* (New York: North Point Press, 2004), 213.

6. Ibid., 202. The Dalton Highway runs approximately five hundred miles north–south from Fairbanks to the Prudhoe Bay oilfields; the Trans-Alaska pipeline runs some eight hundred miles, bisecting the state from Prudhoe Bay to Valdez.

7. Banerjee discusses his model of inhabited practice in his essay "Land as Home: A Portrait of the Circumpolar Arctic in a Rapidly Changing Planet," author's personal website, 2007, accessed April 16, 2013, http://www.subhankarbanerjee.org/banerjee.html. Banerjee first brought up the concept of "ecocultural rights" in his plenary address at ASLE 2010 in Bloomington, Indiana.

8. The access to and protection of "human rights" has been at the forefront of debates about drilling for oil in the Arctic National Wildlife Refuge. Those who are for drilling suggest that economic security in an oil-dependent economy is a human right while those against drilling insist on recognizing as a human right the interdependency of humans, animals, and a healthy environment. Tim Mowry, "Opinions Mixed on Arctic National Wildlife Refuge Designation," *Fairbanks Daily News Miner*, May 14, 2010, http://www.newsminer.com/news/local_news/article_5c76e8cd-622e-531e-9825–4fc8e8fe262c.html?TNNoMobile.

9. I happened to be visiting the community when the polar bear was shot. Part of her story includes her body on display in the hunter's home. While this action may seem to be the ultimate expression of domination, the puzzled and troubled questions that floated around her head pointed to other sentiments that connected the hunter with the bear and the community with federal wildlife laws. Adding to the bear's ambiguity, around this time in Canada's Northwest Territory, an Inuvaluit hunter killed a second-generation hybrid—the offspring of a polar bear–grizzly

hybrid and a grizzly bear. "Bear Shot in N.W.T. Was Grizzly-Polar Hybrid: Could Be First 2nd Generation Hybrid Found in Wild," *CBC*, April 30, 2010, http://www.cbc.ca/news/canada/north/bear-shot-in-n-w-t-was-grizzly-polar-hybrid-1.870506.

10. The Brooks Range is part of the continental divide that separates the flat, "barren-ground" coastal plain with north-running rivers from the black spruce–forested interior with rivers that run south and west.

11. See Berger for a complete discussion of the use of terms such as *Eskimo* and *Indian* to refer to the Native peoples of Alaska. These are, in essence, colonial and modern politicolegal descriptions, not what the people so designated prefer to call themselves. Thomas R. Berger, *Village Journey: The Report of the Alaska Native Review Commission* (New York: Hill and Wang, 1985).

12. The Gwich'in are categorically against it while the Iñupiat favor drilling in the refuge but oppose offshore drilling.

13. Julie Cruikshank, *The Social Life of Stories: Narrative and Knowledge in the Yukon Territory* (Lincoln: University of Nebraska Press, 1998), 111.

14. Moses Peter, "The Ice Bear," trans. Moses P. Gabriel, ed. Craig Mishler (Fairbanks: Alaska Native Language Archive, 1972), 8–9.

15. Barry Lopez, *Arctic Dreams: Imagination and Desire in a Northern Landscape* (New York: Scribner, 1986), 63.

16. Peter, "Ice Bear," 18–19.

17. Katherine Peter, *Vasaagihdzak, Shoh Deetrya'haa Gwandak Tr'injaa* [Three Stories] (Fairbanks: Alaska Native Language Center, 1975), n.p.

18. Nastassia Martin, personal communication with author, February 21, 2010. Martin, an anthropologist, heard Gwich'in traditional chief, Episcopal priest, and cultural activist Trimble Gilbert speak when she visited the community of Arctic Village during the days following the polar bear incident.

19. Robert A. Brightman, *Grateful Prey: Rock Cree Human-Animal Relationships* (Los Angeles: University of California Press, 1993), 205. Brightman explains that Cree "refers today to a continuum of culturally and linguistically related native people inhabiting the subarctic boreal forest from Quebec on the east to British Columbia on the west" (4). While Cree speak a dialect of Algonquin and have had some historical disputes with Athabascan speakers living north and west to their traditional homelands, certain cultural features unite these boreal inhabitants, including the trickster hero of a long story cycle—Vasaagihdzak in Gwich'in Athabascan or Wisahkichak in Cree Algonquin.

20. Institute for Tribal Environmental Professionals, "Tribal Profiles Alaska-Athabascan Region," Nau.edu, December 2, 2012, http://www4.nau.edu/tribalclimatechange/tribes/ak_athabascan.asp.

21. The motivation of the project is "So the Polar Bear Always Has a Place to Call Home." Furthermore, "Together with WWF and people like you, Coca-Cola is helping to establish a place where polar bears and people can thrive in the Arctic." "Arctic Efforts: Creating an Arctic Home," Coca-Cola Arctic Home, Coca-Cola and World Wildlife Fund, accessed April 10, 2013, http://www.arctichome.com.

22. Institute for Tribal Environmental Professionals, "Tribal Profiles Alaska-Athabascan Region."

23. Donna Haraway, "*When Species Meet*: Staying with the Trouble," *Environment and Planning D: Society and Space* 28 (2010): 53–55.

24. Jacques Derrida, *Specters of Marx* (London: Routledge, 1992), xviii–xix.

25. Anthony Fischbach and Catherine Puckett, "Polar Bear Denning Shifting From Sea Ice to Coastal Habitats in Northern Alaska," USGS, July 12, 2007, http://www.usgs.gov/newsroom/article.asp?ID=1705.

26. Ibid.

27. *OED Online*, s.v. "Polar bear" (2nd def.).

28. Lopez, *Arctic Dreams*, 79.

29. For further discussion of animals as projections for human desires and as foils for human agendas—especially in the maintenance of category boundaries around gender, class, and race—cf. Freccero (2011); Haraway (2010); Huggan and Tiffin (2010); Hird (2008); Chris (2006).

30. Subhankar Banerjee, *Arctic Voices: Resistance at the Tipping Point* (New York: Seven Stories Press, 2012).

31. Kirsten Hastrup, "Ultima Thule: Anthropology and the Call of the Unknown," *Journal of the Royal Anthropological Institute* 13 (2007): 789–804.

Bibliography

Atwood, Margaret. *The Tent*. New York: Doubleday, 2006.

Banerjee, Subhankar. *Arctic Voices: Resistance at the Tipping Point*. New York: Seven Stories Press, 2012.

———. "Ecocultural Rights: Stories from the North, and also the South." Lecture for the Association for the Study of Literature and the Environment. University of Indiana, Bloomington, June 25, 2010.

———. "Land as Home: A Portrait of the Circumpolar Arctic in a Rapidly Changing Planet." SubhankarBanerjee.com, 2007, accessed April 16, 2013. http://www.subhankarbanerjee.org/banerjee.html.

"Bear Shot in N.W.T. Was Grizzly-Polar Hybrid: Could Be First 2nd Generation Hybrid Found in Wild." *CBC*, April 30, 2010. http://www.cbc.ca/news/canada/north/bear-shot-in-n-w-t-was-grizzly-polar-hybrid-1.870506.

Berger, Thomas R. *Village Journey: The Report of the Alaska Native Review Commission*. New York: Hill and Wang, 1985.

Brightman, Robert A. *Grateful Prey: Rock Cree Human-Animal Relationships*. Los Angeles: University of California Press, 1993.

Chris, Cynthia. *Watching Wildlife*. Minneapolis: University of Minnesota Press, 2006.

Coca-Cola and World Wildlife Fund. "Arctic Efforts: Creating an Arctic Home." Coca-Cola Arctic Home. Accessed April 10, 2013. http://www.arctichome.com (site discontinued).

Cruikshank, Julie. *The Social Life of Stories: Narrative and Knowledge in the Yukon Territory*. Lincoln: University of Nebraska Press, 1998.

Crutzen, Paul J., and Eugene F. Stoermer. "The Anthropocene." *IGBP Newsletter* 41 (2000): 17.

Derrida, Jacques. *Specters of Marx: The State of the Debt, the Work of Mourning and the New International*. London: Routledge, 2006.

Fischbach, Anthony, and Catherine Puckett. "USGS Finds Polar Bear Denning Shifting From Sea Ice to Coastal Habitats in Northern Alaska." Press release, July 12, 2007. https://www.doi.gov/news/archive/07_News_Releases/070712.html.

Freccero, Carla. "Carnivorous Virility; or, Becoming-Dog." *Social Text* 29, no. 1 106 (2011): 177–195.

"Great Male Polar Bear Swimming in Freezing Seas." From *Planet Earth*, BBC Wildlife: Collectors Edition. September 23, 2011. http://www.youtube.com/watch?v=6AowduxKtN4.

Haraway, Donna. *When Species Meet*. Minneapolis: University of Minnesota Press, 2007.

———. "*When Species Meet*: Staying with the Trouble." *Environment and Planning D: Society and Space* 28 (2010): 53–55.

Hastrup, Kirsten. "Ultima Thule: Anthropology and the Call of the Unknown." *Journal of the Royal Anthropological Institute* 13 (2007): 789–804.

Hird, Myra J. "Animal Trans." *Queering the Non/Human*, edited by Noreen Giffney and Myra J. Hird. Burlington, VT: Ashgate, 2008.

Huggan, Graham, and Helen Tiffin. *Postcolonial Ecocriticism: Literature, Animals, Environment*. London: Routledge, 2010.

"Ice Worlds." *Planet Earth: As You've Never Seen It Before: the Complete Series*. Produced by Alastair Fothergill. 2005; London: BBC, 2006. DVD.

Institute for Tribal Environmental Professionals. "Tribal Profiles Alaska—Athabascan Region." Northern Arizona University. Accessed December 2, 2012. http://www4.nau.edu/tribalclimatechange/tribes/ak_athabascan.asp.

Lopez, Barry. *Arctic Dreams: Imagination and Desire in a Northern Landscape*. New York: Scribner, 1986.

Mowry, Tim. "Opinions Mixed on Arctic National Wildlife Refuge Designation." *Fairbanks Daily News-Miner*. May 14, 2010. http://www.newsminer.com/news/local_news/article_5c76e8cd-622e-531e-9825–4fc8e8fe262c.html?TNNoMobile.

Peter, Katherine. *Vasaagihdzak, Shoh Deetrya'haa Gwandak Tr'injaa* [Three Stories]. Fairbanks: Alaska Native Language Center, 1975.

Peter, Moses. "The Ice Bear." Translated by Moses P. Gabriel. Edited by Craig Mishler. Fairbanks: Alaska Native Language Archive, 1972.

U.S. Fish and Wildlife Service. "Arctic National Wildlife Refuge." Accessed April 10, 2013. http://www.fws.gov/alaska/nwr/arctic/.

Wohlforth, Charles. *The Whale and the Supercomputer: On the Northern Front of Climate Change*. New York: North Point Press, 2004.

PART II

thinking with northern animals

4

Indigeneity and Ecology in Iñupiaq and Faroese Whaling

russell fielding / UNIVERSITY OF THE SOUTH

Introduction

To witness the capture and slaughter of a large marine mammal can be a disturbing experience, visually and psychologically unnerving to many. The process involves shedding an enormous amount of blood, as the nineteenth-century French historian Jules Michelet[1] noted: "the Whale, when wounded, ensanguines the ocean to a great distance; the blood that we have in drops, is lavished upon him in torrents." While this bloodshed can also be seen as an indicator of a quick death for the whale, the incongruous and unexpected sight of a reddened harbor, beach, ice floe, or seascape is often the photographic subject of advertisements calling for action against whaling nations.[2] "If only the blood was blue," the Faroese Minister of Fisheries, Bjørn Kalsø, lamented when I interviewed him in 2005.

At the same time, for a community that practices whaling, the death of a whale signifies the beginning of a time of plenty. Some celebrate; others work to exhaustion; many lend a hand in labor and nearly all know that they will eat well. The blood, if it is considered at all, is seen as a necessary part of the act of food production, not as evidence of a crime against nature.

This chapter intervenes in this volume's consideration of charismatic megafauna as metonyms of the North by considering two whaling communities, the Iñupiat and the Faroese, along with their traditions and the reactions they provoke. Both communities are based in the North and practice whaling as a means of food production. After introducing the communities and their whaling practices, I discuss the international regulation of whaling with a particular focus on what is termed "aboriginal subsistence whaling." Following this, I look more closely at the concept of indigeneity, that is, the state of being indigenous. Indigeneity can be reckoned for a variety of reasons in many contexts but with regard to having one's whaling practices accepted by those outside one's own culture, indigeneity is essential.

Iñupiaq Whaling

In their "Taxonomy of World Whaling," Randall Reeves and Tim Smith[3] place the origin of "arctic aboriginal" whaling at least two thousand years before the present. As Reeves and Smith acknowledge, arctic whaling has always been both geographically discontinuous and transnational. Within the broad context of arctic whaling, this essay focuses on the Iñupiat of Alaska's North Slope Borough. This group refers to itself as the "People of the Whales" and, as Chie Sakakibara has argued, derives much of its cultural identity from "the whaling cycle," or the system of subsistence and cultural activities, within which whaling is central.[4] The primary target of this hunt is the bowhead whale (*Balaena mysticetus*). Owing to the sheer size of the bowhead, the hunting, hauling ashore, butchering, and consumption of just one whale is necessarily a community event. The community nature of Iñupiaq whaling has served to maintain its sustainability by ensuring that food products derived from the hunt are distributed, rather than hoarded, and by disincentivizing the taking of excess whales.

Faroese Whaling

Inhabitants of the Faroe Islands, a Danish archipelago in the North Atlantic Ocean, hunt long-finned pilot whales (*Globicephala melas*), using a method known as drive-style whaling (called *grindadráp* in Faroese, often shortened to *grind*). This involves coordinated efforts to drive a pod of whales into a fjord, where they will beach or become stranded in the shallow water. Once stranded, the whales are hauled ashore with hooks and ropes and are killed by shore-based whalers using small handheld knives. Hunts are ad hoc and unpredictable, occurring only when whales are sighted within range of the islands. The first recorded instance of this activity occurred in 1587 but most historians trace its introduction to the time of Norse settlement in the ninth century.[5]

Faroese law has codified several conservation strategies that have been in place for many centuries. The most visible and effective is the geographical limitation of whaling activities to certain approved whaling beaches. Having a roster of approved whaling beaches allows the authorities to allow or disallow any impending whale drive. Drives have been disallowed if weather or ocean conditions are unfavorable or if the food that would result is not needed.

Regulations, Exceptions, and Ambiguities

In 1946, fifteen whaling nations met to create the International Whaling Commission (IWC)—the first and only worldwide regulatory body for whaling.[6] In 1972, nonwhaling nations began joining the IWC, "apparently for the specific purpose of changing the balance of votes and thus to institute a moratorium policy."[7] This apparent goal was met in 1982, when the IWC passed a resolution that would reduce the annual quota for commercial operations targeting all whale species to zero, beginning in 1986.

Great Whales vs. Small Cetaceans

In the establishment of the IWC moratorium, a de facto exception was made for the hunting of "small cetaceans," which include any whale or dolphin besides the twelve species of baleen whales and the sperm whale (*Physeter macrocephalus*). These thirteen species are collectively known as the "great whales" or "IWC whales." The IWC's competence for regulation is generally seen as only extending to these thirteen species.

This distinction is problematic to some because the categories of great whales (which are protected by the IWC) and small cetaceans (which are not) are not determined on the basis of conservation status or, as Alexander Gillespie[8] points out, even body size. Rather, the twelve species of baleen whale and the sperm whale are those that were hunted most extensively during the most recent period of commercial whaling.

Additionally, Arne Kalland has shown that, in terms used in antiwhaling literature, species and even higher-level taxonomic distinctions are intentionally blurred. Thus we have a more developed history of the "save the whales" movement than of specific efforts to protect certain threatened or endangered whale or dolphin species. This failure to recognize the different conservation needs of various cetacean species has led to the insistence that whales are endangered.[9] Naturally it would follow that, if all whales are indeed endangered, it should be only under very special circumstances that humans be allowed to kill any whale. The most enduring, and the most widely accepted, of such special circumstances is aboriginal subsistence whaling. However, defining this type of whaling—indeed, defining any type of whaling—has historically been difficult and inexact.

Commercial vs. Subsistence

From its outset, the moratorium was focused on "commercial" whaling, as indicated by the text itself: "catch limits for the killing for commercial purposes of whales from all stocks . . . shall be zero."[10] Exceptions were made for scientific whaling and for "subsistence whaling by aborigines."[11] These types of whaling are not subject to the moratorium, though the IWC can impose limits on the number of whales caught by specific groups. There are currently "aboriginal" groups in four countries that take whales legally under the IWC's oversight: Denmark (specifically Greenland), Russia (specifically Siberia), St. Vincent and the Grenadines (specifically the island of Bequia), and the United States (specifically the states of Alaska and Washington). The diversity contained within this roster of whaling nations indicates the imprecision associated with the definition of "aboriginal" or "indigenous."

In her essay on environmental justice and the regulation of whaling, Gail Osherenko provides a salient example of this imprecision through her use of an array of diverse descriptors for the various whaling operations she discusses.[12] Osherenko introduces Faroese whaling as "a nonnative harvest of pilot whales" in contrast with "native hunts" in Canada and "aboriginal hunting in Greenland." She goes on to note that "the United States has supported the rights of aboriginal Iñupiat and Makah whalers" while mounting pressure upon Japanese and other "traditional coastal whaling communities."[13] Elsewhere Osherenko refers to "aboriginal subsistence whaling," "artisanal whaling," "commercial whaling," and "industrial whaling." These quick shifts among these terms indicate the complicated nature of attempting to categorize whaling activities as being of one type or another.

Another criterion used is the economic purpose of whaling, whether for food or for profit. However, as several scholars have pointed out,[14] there is significant overlap and mingling between the two. Indeed, the two locations in this study provide a salient example. Iñupiat give, barter, and sell whale products within and among communities.[15] In the Faroe Islands the sale of whale products is officially forbidden—although unofficial sales do take place.[16]

While terms such as *commercial* and *subsistence* may not offer much concrete substance to inform our discussion of indigeneity in whaling, it would seem more productive to ask whether or not those doing the whaling are, ethnically, indigenous peoples. However, as we shall see, the answer to this question is also difficult to pin down as it is laden with assumptions that bear marks of racist, classist, and colonial thought.

Indigeneity and Whaling

Indigenous people have been defined by Anaya as "the living descendants of preinvasion inhabitants of lands now dominated by others."[17] This definition—developed for purposes of international law—diverges from the implicit definition used by the IWC. For example, the Caribbean island of Bequia holds an aboriginal whaling permit from the IWC that allows the take of up to four humpback whales (*Megaptera novaeangliae*) per year. Ethnically, Bequians descend primarily from African slaves and British colonialists. Few, if any, claim ancestral links to indigenous Caribbean groups such as the Arawaks and Caribs, who, incidentally, are not known to have hunted marine mammals.[18] Bequian whaling history is traced to the nineteenth-century "Yankee whalers" from the United States. As whale stocks diminished, it became less profitable for American whalers to make the long trip from their home ports in New England. The niche they left open was soon filled by local Bequians who had learned the skills of whaling aboard the American vessels. It was only during the early twentieth century that local whaling efforts became well established on Bequia and throughout the region.[19]

The Bequia example indicates that indigeneity, as defined by Anaya, is not essential in the acquisition of IWC permission for aboriginal subsistence whaling. However, as we have seen, IWC definitions (e.g., great whales vs. small cetaceans) do not always equate to public perceptions. The cultural histories of the Iñupiat and the Faroese further complicate this lens of indigeneity.

Iñupiaq History

The Iñupiat trace their history to the Thule people, themselves the second major wave of immigrants who crossed the Bering Land Bridge and spread across the northern part of North America including present-day Alaska, Canada, and Greenland. Whaling has been a major food-production strategy for the Thule and their cultural descendants, including the Iñupiat, for centuries.[20] Climate change and contact with Europeans brought further adaptation, including trade and the introduction of a cash economy and new foodstuffs to the Iñupiaq diet. Guns and eventually snowmobiles were also acquired, both of which the Iñupiat integrated into their hunting methods.

In 1867, when the United States purchased Alaska from Russia, and again in 1959, when Alaska gained statehood, land ownership among all Alaska Native groups was a contentious issue. The Alaska Native Claims Settlement Act of 1971 addressed the issue by creating twelve regional corporations as landholders of traditional native lands.[21] In the North Slope Borough, the Ukpeaġvik Iñupiat Corporation, based in Barrow, is the major landholder under the act. Today the Iñupiat are recognized by the U.S. government and are given limited self-government through the Iñupiat

Community of the Arctic Slope. Their status within the United States arguably fits Anaya's definition of indigenous people.

Faroese History

Modern Faroese descend from the Norse settlers who came to the islands in the late ninth century in what Carl Sauer called a "great Viking exodus."[22] Recently it has been shown that these Norse settlers were not the first inhabitants of the Faroe Islands.[23] Research is currently under way to determine who, exactly, these pre-Norse inhabitants were, but evidence from place-names, literature, and folklore points to a colony of Irish monastics.[24] Considering Anaya's definition of indigenous peoples, could the Faroese be thought of as the "others" who currently dominate the land? Perhaps, though the "preinvasion inhabitants" have either long since moved on, been eradicated, or integrated into the Faroese population.

Since 1948, Denmark has allowed the Faroese government home rule but maintains sovereignty, specifically in the areas of defense and international relations. Would this relationship define the Faroes as "lands now dominated by others"? While few would define the Faroese–Danish relationship as being based on "domination," the Faroese are still not fully autonomous. However, as Godfrey Baldacchino reminds us, the relationship between the Faroes and Denmark is "dissolvable by mutual consent."[25] Multiple referenda have yet to succeed in dissolving the bond between the Faroe Islands and Denmark. These failed independence referenda indicate that the majority of Faroese people may not see themselves as being dominated. As for whaling, they are free to continue on their own terms. According to an adviser on Faroese matters at the Danish Ministry of Foreign Affairs, Denmark defers all decisions on whaling policy to the Faroese Home Rule government.[26]

Still, the case of Bequia calls out for attention. By whom are the Bequians "dominated"? Certainly the British once held the Grenadines as a colony but that structure dissolved in 1979. St. Vincent and the Grenadines is an independent country. If the Bequians, with their hundred-year history of whaling, can qualify for an aboriginal subsistence permit from the IWC, could not the Faroese justify their own "thousand-year"[27] take of small cetaceans before a court of public opinion?

Indigenous Need

Both the Iñupiat and the Faroese have long histories of whaling in the places where they currently reside. Whaling has become entwined within the cultures of both groups and continues to provide food in both contexts. According to the IWC, one objective of allowing indigenous whaling is to "enable native people to hunt whales at levels appropriate to their cultural and nutritional requirements (also called 'need')."[28] Thus we consider the question of need, along with the existence—and relevance—of other alternative methods of food production available to whaling communities.

Under "Position on Aboriginal Whaling," Sea Shepherd writes that it opposes

> all whaling by any people, anywhere for any reason. However, our activities are directed at illegal whaling operations. The hunting of endangered bowheads by Northern aboriginal people is legal according to the International Whaling Commission. Therefore, there has been no opposition to these practices by Sea Shepherd Conservation Society.[29]

Sea Shepherd draws a firm distinction between whaling by "Northern aboriginal people" and Faroese whaling—to which it refers using the Faroese term *grind*—noting that the organization stands

> against the slaughter of Pilot Whales in the Danish Faeroe Islands. Sea Shepherd intends to intervene in the *grind*, when at all possible, to prevent the unnecessary loss of such precious marine wildlife that has no place in modern times.[30]

Much of the opposition to Faroese whaling, as recorded meticulously and critically by Kate Sanderson, points out the believed incongruity between a European society with a "modern" lifestyle and an ongoing reliance upon whaling for food. Sanderson cites an opponent as saying, in reference to the Faroese, "if they want to kill whales in the traditional way, that's fine by us, if nothing else about their way of life, significantly anyway, has changed."[31] Sanderson touches on issues of ethnicity and modernity with her discussion of the "ambiguity" of "the sight of fair-skinned men, fully clothed in woolen sweaters . . . entering the water with knives in their hands to kill whales."[32] She goes on to question the "simplistic urban view" held by opponents of Faroese whaling, who rely on a stereotype of what subsistence hunting should look like: "the hunter is at one with the unspoiled wilderness in which he hunts and must not therefore display any of the incongruous trappings or influences of 'modern civilization.'"[33] In the discourse of antiwhaling that Sanderson critiques, one must be either fully modern or fully engaged in subsistence activities. The reference to "fair-skinned men, fully clothed" reveals a barely subliminal ethnic divide along this overtly economic distinction as well. In this discourse, facets of Faroese culture such as the publicly viewable killing of marine mammals, the free distribution of the food products that the activity produces, and of course the harbor waters stained red with the blood of whales simply do not fit with the motorboats, automobiles, and Scandinavian standard of living enjoyed by these ethnically European islanders.

According to the Whale and Dolphin Conservation Society, a London-based animal welfare organization,

> Pilot whales have been killed in the Faroe Islands for over 400 years and it can be appreciated that in the past they made an important contribution to the Islanders' survival in the harsh conditions of these isolated islands. . . . Today, however, the Faroese people enjoy a standard of living at least as high as that of any other country in Scandinavia. The consumption of pilot whales is, therefore, thought to be no longer necessary for survival.[34]

This puts to words an oft-unspoken assumption in the argument against whaling—that whales should be a food source of last resort. Paul Watson of Sea Shepherd speaks for many who oppose Faroese whaling when he calls it "unjustified" owing to it being "not necessary for subsistence" and goes on to say that "you cannot compare the materially wealthy communities in the Faeroes with subsistence communities."[35] Why, though, should whaling be conducted only by a people without other options? Could it truly be said that any people "need" any one particular foodstuff? Perhaps. Certain foods are enmeshed within their cultures to the point that one could hardly conceive of one without the other.[36] However, these enmeshments are primarily, if not entirely, cultural. Contextually, the references to "need" by those opposed to whaling primarily indicate the consideration of nutritional, not cultural, need.

The Question of Conservation

We have seen that the effort to draw clear distinctions among whaling operations is beset by a host of assumptions, inconsistencies, and self-contradictions. In the social aspect, elements of colonialism and racism lie just beneath the surface. Biologically, the discourse sets no clear taxonomic line between *great whales* and *small cetaceans*. Economic terms like *commercial*, *subsistence*, and *artisanal* are used without precision. Even the term *indigenous* is applied inconsistently within the whaling debate, as the case of Bequia shows. To ask whether a group "needs" to whale is to invite a subjective answer that judges the availability and suitability of various alternative foodstuffs, attempts to decide for the local people which alternatives should suffice, and ordinarily ignores the "cultural . . . requirements"[37] that the IWC ostensibly seeks to protect through its policy on aboriginal subsistence whaling. Perhaps another measure of indigeneity—a conservation-centered measure—would be fit to consider. To do so would be to shift the whaling regime toward an evidence-based exception process by which those groups that wish to have their whaling activities sanctioned by the IWC would show their conservation methods to be based upon long-standing cultural tradition. It would avoid the assumption of the "ecological Indian" described by Shepard Krech and expounded upon by Sarah Jaquette Ray, by which indigenous subsistence activities are deemed sustainable based simply on the ethnicity of the persons conducting them.[38] Rather, it would remove ethnicity from the decision and instead focus on indigeneity—not of the whalers but of the conservation methods they employ.

The application of an indigenous-conservation criterion to the two cultures under consideration here reveals the existence of deeply ingrained conservation strategies such as the geographical limitation of whaling bays in the Faroe Islands and the community involvement in Iñupiaq whaling. Such practices, enforced by the weight of tradition, are arguably more likely to endure and to maintain public adherence than conservation regimes that are based upon the rule of law alone.[39]

A conservation-centered approach to indigenous whaling would enact moratoria on any operation targeting whales of an endangered species. It would allow limited whaling to be conducted by groups who have effective conservation methods ingrained within their cultural traditions, with the understanding that the killing or saving of a animal need not always be mutually exclusive (Boyer, chapter 5). This approach would take into account the regimes of measured resource use that have evolved—whether consciously or not—to effect conservation of whale stocks. If managed through an organization like the IWC, it would build upon the statements of "cultural and nutritional requirements,"[40] asking the group to put forth a statement explaining its own historical management strategies. It would seek to balance the needs of the human population with the stability of the whale population.

Conclusion

The suggestion above is but one possible way forward that would offer some clarity to the whaling debate. People who oppose whaling do so for a variety of reasons. Some, from an animal-rights perspective, are against all forms of hunting. Others feel that whales should be given special consideration, owing to their intelligence, size, or other attributes. However, the IWC was founded to conserve whale stocks. A conservation-centered approach to whaling would align more accurately with the organization's mission than the current multifaceted parsing of whaling operations based upon the target species and the cultural and political history of the people involved. A conservation-centered management regime for whaling would question not the indigeneity of the whalers but rather the indigeneity of the methods of conservation in use.

Endnotes

1. Jules Michelet, *La Mer* (New York: Rudd and Carleton, 1861), 229.

2. For an analysis of these media, see Kate Sanderson, "*Grind*—Ambiguity and Pressure to Conform: Faroese Whaling and the Anti-Whaling Protest," in *Elephants and Whales: Resources for Whom?*, eds. Milton Freeman and Urs P. Kreuter (Basel, Switzerland: Gordon and Breach, 1994), 187–201.

3. Randall Reeves and Tim Smith, "A Taxonomy of World Whaling: Operations and Eras," in *Whales, Whaling, and Ocean Ecosystems*, ed. J. A. Estes et al. (Berkeley: University of California Press, 2006), 88.

4. Chie Sakakibara, "*Kiavallakkikput Agviq* (Into the Whaling Cycle): Cetaceousness and Climate Change Among the Iñupiat of Arctic Alaska," *Annals of the Association of American Geographers* 100, no. 4 (2010).

5. Jonathan Wylie, "Grindadráp," in *The Ring of Dancers: Images of Faroese Culture*, eds. Jonathan Wylie and David Margolis (Philadelphia: University of Pennsylvania Press, 1981); Arne Thorsteinsson, "Hvussu Gamalt er Grindadrápið?" *Varðin* 53 (1986).

6. Ray Gambell, "International Management of Whales and Whaling: An Historical Review of the Regulation of Commercial and Aboriginal Subsistence Whaling," *Arctic* 46, no. 2 (1993).

7. Harry N. Scheiber, "Historical Memory, Cultural Claims, and Environmental Ethics in the Jurisprudence of Whaling Regulation," *Ocean & Coastal Management* 38 (1998): 14.

8. "Small Cetaceans, International Law and the International Whaling Commission," *Melbourne Journal of International Law* 2, no. 2 (2001).

9. Arne Kalland, "Management by Totemization: Whale Symbolism and the Anti-Whaling Campaign," *Arctic* 46, no. 2 (1993).

10. International Whaling Commission, *International Convention for the Regulation of Whaling, 1946: Schedule, as amended by the Commission at the 64th Annual Meeting Panama City, Panama, July 2012* (Cambridge, UK: IWC, 2012), 5.

11. International Whaling Commission, *Report of the special working group of the Technical Committee concerning management principles and development of guidelines about whaling for subsistence by aborigines* (Cambridge, UK: IWC, Document IWC/33/14, 1981).

12. Gail Osherenko, "Environmental Justice and the International Whaling Commission: *Moby-Dick* Revisited," *Journal of International Wildlife Law & Policy* 8, no. 2–3 (2005): 226, footnote 28 (emphasis added).

13. Osherenko, "Environmental Justice," 228 (emphasis added).

14. For exampe, Brian Moeran, "The Cultural Construction of Value: Subsistence, Commercial, and Other Terms in the Debate about Whaling," *Maritime Anthropological*

Studies 5, no. 2 (1992); Randall R. Reeves, "The Origins and Character of 'Aboriginal Subsistence' Whaling: A Global Review," *Mammal Review* 32, no. 2 (2002).

15. Milton M. R. Freeman et al., *Inuit, Whaling, and Sustainability* (Lanham, MD: AltaMira Press, 1998).

16. Russell Fielding, "Artisanal Whaling in the Atlantic: A Comparative Study of Culture, Conflict, and Conservation in St. Vincent and the Faroe Islands" (PhD diss., Louisiana State University, 2010).

17. S. James Anaya, *Indigenous Peoples in International Law* (New York: Oxford University Press, 1996), 3.

18. Reeves, "Origins and Character," 2002.

19. John E. Adams, "Historical Geography of Whaling in Bequia Island, West Indies," *Caribbean Studies* 11 (1971); Nathalie Ward, *Blows, Mon, Blows! A History of Bequia Whaling* (Woods Hole, MA: Gecko Productions, 1995).

20. Reeves, "Origins and Character"; Reeves and Smith, "A Taxonomy of World Whaling"; Bruce G. Trigger and Wilcomb E. Washburn, *The Cambridge History of the Native Peoples of the Americas* (Cambridge, UK: Cambridge University Press, 1996).

21. Ukpeaġvik Iñupiat Corporation, "Alaska Native Claims Settlement Act (ANCSA)," 2011, accessed December 8, 2013, http://www.ukpik.com/ANCSA.htm (site discontinued).

22. Carl Sauer, *Northern Mists* (Berkeley: University of California Press, 1968), 85.

23. Mike J. Church et al., "The Vikings Were Not the First Colonizers of the Faroe Islands," *Quaternary Science Reviews* 77 (2013).

24. Sauer, *Northern Mists*.

25. Godfrey Baldacchino, *Island Enclaves: Offshoring Strategies, Creative Governance, and Subnational Island Jurisdictions* (Montreal: McGill-Queen's University Press, 2010), 55.

26. Árni Olafsson, e-mail message to author, February 24, 2010.

27. Séan Kerins, *A Thousand Years of Whaling: A Faroese Common Property Regime* (Edmonton: CCI Press, 2010).

28. International Whaling Commission, "Aboriginal Subsistence Whaling," accessed December 8, 2013, http://iwc.int/aboriginal.

29. Sea Shepherd, "Whaling Around the World," accessed December 6, 2013, http://www.seashepherd.org/whales/whaling-around-the-world.html.

30. Sea Shepherd, "Operation Ferocious Isles—Sea Shepherd's Pilot Whale Defense Campaign in the Faeroe Islands," accessed December 6, 2013, http://www.seashepherd.org/ferocious-isles/ferocious-isles.html.

31. Sanderson, "*Grind*—Ambiguity and Pressure to Conform," 187.

32. Ibid., 198.

33. Ibid.

34. Phillippa Brakes et al., *Troubled Waters: A Review of the Welfare Implications of Modern Whaling Activities* (London: World Society for the Protection of Animals, 2004).

35. Paul Watson, e-mail message to author, October 25, 2005.

36. Richard Wilk, "'Real Belizean Food': Building Local Identity in the Transnational Caribbean," *American Anthropologist* 101 (1999); Igor Cusack, "African Cuisines: Recipes for Nation-building?" *Journal of African Cultural Studies* 13 (2000); Pauliina Raento, "Changing Food Culture and Identity in Finland," in *Finnishness in Finland and North America: Constituents, Changes, and Challenges*, ed. Pauliina Raento (Toronto: Aspasia Books, 2006).

37. IWC, "Aboriginal Subsistence Whaling," 2013.

38. Shepard Krech III, *The Ecological Indian: History and Myth* (New York: W.W. Norton, 2000); Sarah Jaquette Ray, *The Ecological Other: Environmental Exclusion in American Culture* (Tucson: University of Arizona Press, 2013), ch. 2.

39. Nancy C. Doubleday, "Aboriginal Subsistance [*sic*] Whaling: The Right of Inuit to Hunt Whales and Implications for International Environmental Law," *Denver Journal of International Law and Policy* 17, no. 2 (1989).

40. Ibid.

Bibliography

Adams, John E. "Historical Geography of Whaling in Bequia Island, West Indies." *Caribbean Studies* 11 (1971): 55–74.

Anaya, S. James. *Indigenous Peoples in International Law*. New York: Oxford University Press, 1996.

Baldacchino, Godfrey. *Island Enclaves: Offshoring Strategies, Creative Governance, and Subnational Island Jurisdictions*. Montreal: McGill-Queen's Press University, 2010.

Brakes, Phillippa, et al. *Troubled Waters: A Review of the Welfare Implications of Modern Whaling Activities*. London: World Society for the Protection of Animals, 2004.

Church, Mike J., et al. "The Vikings Were Not the First Colonizers of the Faroe Islands." *Quaternary Science Reviews* 77 (2013): 228–232.

Cusack, Igor. "African Cuisines: Recipes for Nation-building?" *Journal of African Cultural Studies* 13 (2000): 207–225.

Doubleday, Nancy C. "Aboriginal Subsistance [*sic*] Whaling: The Right of Inuit to Hunt Whales and Implications for International Environmental Law." *Denver Journal of International Law and Policy* 17, no. 2 (1989): 373–393.

Fielding, Russell. "Artisanal Whaling in the Atlantic: A Comparative Study of Culture, Conflict, and Conservation in St. Vincent and the Faroe Islands." PhD dissertation, Louisiana State University, 2010.

Freeman, Milton M. R., et al., eds. *Inuit, Whaling, and Sustainability*. Lanham, MD: Altamira Press, 1998.

Gambell, Ray. "International Management of Whales and Whaling: An Historical Review of the Regulation of Commercial and Aboriginal Subsistence Whaling." *Arctic* 46, no. 2 (1993).

Gillespie, Alexander. "Small Cetaceans, International Law and the International Whaling Commission." *Melbourne Journal of International Law* 2 (2001): n.p.

International Whaling Commission. "Aboriginal Subsistence Whaling." Accessed December 8, 2013. http://iwc.int/aboriginal.

———. *International Convention for the Regulation of Whaling, 1946: Schedule, as amended by the Commission at the 64th Annual Meeting Panama City, Panama, July 2012*. Cambridge, UK: IWC, 2012.

———. *Report of the special working group of the Technical Committee concerning management principles and development of guidelines about whaling for subsistence by aborigines*. Cambridge, UK: IWC, Document IWC/33/14, 1981.

Kalland, Arne. "Management by Totemization: Whale Symbolism and the Anti-Whaling Campaign." *Arctic* 46, no. 2 (1993): 124–133.

Kerins, Séan. *A Thousand Years of Whaling: A Faroese Common Property Regime*. Edmonton: CCI Press, 2010.

Krech, Shepard, III. *The Ecological Indian: History and Myth*. New York: W.W. Norton, 2000.

Michelet, Jules. *La Mer*. New York: Rudd and Carleton, 1861.

Moeran, Brian. "The Cultural Construction of Value: Subsistence, Commercial, and Other Terms in the Debate about Whaling." *Maritime Anthropological Studies* 5, no. 2 (1992): 1–15.

Osherenko, Gail. "Environmental Justice and the International Whaling Commission: *Moby-Dick* Revisited." *Journal of International Wildlife Law & Policy* 8, no. 2–3 (2005): 221–239.

Raento, Pauliina. "Changing Food Culture and Identity in Finland." In *Finnishness in Finland and North America: Constituents, Changes, and Challenges*, 50–71. Toronto: Aspasia Books, 2006.

Ray, Sarah Jaquette. *The Ecological Other: Environmental Exclusion in American Culture*. Tucson: University of Arizona Press, 2013.

Reeves, Randall R. "The Origins and Character of 'Aboriginal Subsistence' Whaling: A Global Review." *Mammal Review* 32, no. 2 (2002): 71–106.

Reeves, Randall R., and Tim D. Smith. "A Taxonomy of World Whaling: Operations and Eras." In *Whales, Whaling, and Ocean Ecosystems*, edited by James A. Estes et al., 82–101. Berkeley: University of California Press, 2006.

Sakakibara, Chie. "*Kiavallakkikput Agviq* (Into the Whaling Cycle): Cetaceousness and Climate Change Among the Iñupiat of Arctic Alaska." *Annals of the Association of American Geographers* 100, no. 4 (2010): 1003–1012.

Sanderson, Kate. "*Grind*—Ambiguity and Pressure to Conform: Faroese Whaling and the Anti-Whaling Protest." In *Elephants and Whales: Resources for Whom?*, edited by Milton Freeman and Urs P. Kreuter, 187–201. Basel, Switzerland: Gordon and Breach, 1994.

Sauer, Carl O. *Northern Mists*. Berkeley: University of California Press, 1968.

Scheiber, Harry N. "Historical Memory, Cultural Claims, and Environmental Ethics in the Jurisprudence of Whaling Regulation." *Ocean & Coastal Management* 38 (1998): 5–40.

Sea Shepherd. "Operation Ferocious Isles—Sea Shepherd's Pilot Whale Defense Campaign in the Faeroe Islands." Accessed December 6, 2013. http://www.seashepherd.org/ferocious-isles/ferocious-isles.html.

———. "Whaling Around the World." Accessed December 6, 2013. http://www.seashepherd.org/whales/whaling-around-the-world.html.

Thorsteinsson, Arne. "Hvussu Gamalt er Grindadrápið?" *Varðin* 53 (1986): 65–66.

Trigger, Bruce G., and Wilcomb E. Washburn. *The Cambridge History of the Native Peoples of the Americas*. Cambridge: Cambridge University Press, 1996.

Ukpeaġvik Iñupiat Corporation. "Alaska Native Claims Settlement Act (ANCSA)." 2011. Accessed December 8, 2013. http://www.ukpik.com/ANCSA.htm.

Ward, Nathalie. *Blows, Mon, Blows! A History of Bequia Whaling*. Woods Hole, MA: Gecko Productions, 1995.

Wilk, Richard. "'Real Belizean Food': Building Local Identity in the Transnational Caribbean." *American Anthropologist* 101 (1999): 244–254.

Wylie, Jonathan. "Grindadráp." In *The Ring of Dancers: Images of Faroese Culture*, edited by Jonathan Wylie and David Margolis, 95–132. Philadelphia: University of Pennsylvania Press, 1981.

5

Saving the Polar Bear and Other Objects

kurtis boyer / LUND UNIVERSITY

Introduction

Animals have played very important roles in the development of human culture and economy. On the basis of being inhuman, we have turned them into everything from symbols of the supernatural to an ingredient in our lipstick. The existence of nonhumans provides a symbolic distinction between *Homo sapiens* and the rest of the animal world. Further, as is explored in this chapter, our encounters and treatment of the nonhuman animal allow for a performative space to *be human*. In particular, this chapter explores how we become a *specific kind* of human based on how we as individual people navigate and maintain our relations with all of the creatures currently defined as nonhuman. To begin, though, a particular subject-object relation must be clarified, where the subjectivity of the nonhuman is flattened to allow for a complete monopoly over what a given encounter with the nonhuman means and should be valued as.

This chapter explores two contrasting models of human-animal relations, represented here by diverging ways humans have come to relate to the polar bear, and ultimately argues that a "third way" is ideal. These two models differ in the way that the subjectivity and moral agency of the nonhuman is conceptualized during a human-animal encounter. The first section provides a point of contrast to the dominant model of human-animal relations that is explored in the rest of the chapter. In this first example, the polar bear is an active participant, and therefore serves a required role in how the meaning and value of the encounter between human and animal is produced. The second section explores the conventional model of human-animal relations, whereby nonhuman subjectivities are ignored in order to appropriate the meaning and value of both the animals and our encounters with them. These processes provide a way for us to be human in a nonhuman world, and subsequently mean that both our killing and saving of the polar bear are emerging from the same set of humancentric subject-object relations. Once we realize this, we can imagine a third way that argues for conservation from a position of relationality, not just aesthetic consumption or, alternatively, subsistence.

The Polar Bear as a Person

As there is no preordained way that people come to relate to nonhuman animals, there is always a possibility that there will be dissenting views on how animals should or can be regulated. An example of this sort of conflict was made evident through what researchers Dowsley and Schmidt came to note as being a state of worry that some hunters in Nunavut expressed following the establishment of hunting quotas. Specifically, the hunters seemed to worry what the polar bear might think of the system and, in consequence, them as hunters. The new quotas were designed to restrict the hunting of bears and were part of a larger science-based wildlife management regime. The hunters believed that such a system would be perceived as arrogant by the polar bear, and thus feared possible retribution.[1] Dowsley and Schmidt provide more detail on the divergent ways of thinking of polar bear agency in the following statement:

> On the one hand, the Inuit perspective emphasizes the qualitative aspects of properly maintaining the bear-human relationship as a key component in authorizing rights since bears give themselves to those hunters and human communities who have the appropriate qualities and maintain proper relationships. On the other hand, the quota system attempts

> to quantify the polar bear hunt by making the contestable assertion that the bears play no role in determining the number of total bears harvested and neither do they participate in supply/demand scenarios or types and methods of consumption.[2]

The tension between the Inuit hunter and the quota system is in part a consequence of diverging orientations during human-animal relations. The hunter thinks of the bear as being able to conceive of and judge his actions. The concerns of the hunter imply that the bear has a theory of mind—that is, that he or she is able to infer meaning from the hunter's actions by reflecting upon and deciding what the hunter may or may not be thinking. Further, it seems that the bear is thought of as practicing a level of moral agency where judgments and actions are made according to a specific set of moral principles that he or she subscribes to. Not only is the polar bear referred to as an individual (as opposed to a species or population), but, with the level of agency it seems to embody, referring to it as a person might be more suitable. This implies that any encounter with the bear must be the result of the bear exercising its own subjecthood and agency. The occurrence of the encounter is dependent on the bear deciding to be party to it. Therefore, within this type of human-animal relations model, the meaning of any given encounter between human and nonhuman must in some way account for both human and nonhuman subjectivities. While it may be presumptuous to think the bear is willing to give itself up for slaughter, the belief that the animal *gives* its body for the "human greater good" still implies that the encounter depends on the bear for its participation. The claim here is not that the subject-object relations described by Dowsley and Schmidt are somehow nonspeciesist or representative of a pan-Inuit human-animal relations model. For the purposes of this chapter, the hunter–polar bear interaction serves as a counter example to the conventional human-animal relations model explored throughout the rest of the chapter. As such, the sorts of subject-object relations that are examined during this chapter should be thought of as only subtle deviations of humancentric ways of thinking of and being with animals—as opposed to being expressive of a particular culture, class, or gender.

The Polar Bear as an Object

Modern species advocacy campaigns and wildlife management frameworks reflect a type of human-animal relations that disregards nonhuman subjectivity and agentic capacity.[3] While modernity has long been credited with the split between the social and the natural, Eckersley notes how it has only recently that scholarly focus has begun to examine the constitutive gap between human and nonhuman communities.[4] Martinelli argues that modern species advocacy and management fall under a "scientific-pragmatic" dimension of anthropocentrism—where animals are thought of as passive objects with lives that can be quantified, measured, and controlled.[5] Within this framework, individual nonhuman subjectivities are ignored and their meaning and value as sentient living beings becomes subsumed by their role in augmenting the lived experience of humans. For example, an individual polar bear, with her own lived experiences, becomes something of a cardboard cutout of herself: a *species*, a *living resource*, or a *national treasure*. This terminological transition, away from being thought of as an individual or groups of individuals, implies an instrumentalizing of the animal's value.[6] As a passive object that is alienable from itself, an animal such as the polar bear becomes fully knowable and appropriable. By thinking of animals not as independent subjectivities but as passive objects, the meaning supplied by our encounters with them is monopolized for a variety of identity-building projects. For example, a *species* is a part of a natural landscape and is thus part of a global (human) natural heritage (often referred to as "the animal kingdom"); a *living resource* is material sustenance for a culture or livelihood; and a *national treasure* supplies symbolic value to a sense of national identity.

In Nunavut, human-animal relations come to be constructed in a variety of ways. In contrast to what was illustrated by the hunter-bear relationship, trophy hunting provides another, entirely distinct version of subject-object relations. As a source of cash revenue, sport hunting and the sale of skins are extremely important for many small, relatively economically undiversified settlements in northern Canada.[7] A steady demand for polar bear trophies has come from wealthy tourist "sport" hunters who are willing to pay up to $30,000 USD for the opportunity to shoot a polar bear and take their trophy home.[8] Hunting tags can be given to either commercial or subsistence hunting enterprises. Local management boards provide communities a say in where the total allowable kills should be distributed. Since the inception of this type of localized management regime in Nunavut, tag allotments for the trophy hunt have increased steadily in absolute terms as well as relative to those allotted to the subsistence hunt.[9] The commodification of wildlife is generally viewed by ecofeminist[10] and critical theory[11] perspectives as being part of the process of distancing and domination that occurs between human and nonhuman communities. However, more than simply institutionalizing a set of relations that turn the bear into a passive object for maximized economic utility, the trophy hunt, and more specifically the defense of it by Inuit leaders, demonstrates the degree to which animal subjectivity has been removed from modern discourses on cultural and economic rights. For example, in response to international opposition to the commercial harvesting of the bear, Inuit leader Mary Simon argued:

> I don't understand, and many Inuit don't understand to this day, why the US is so intent on actively pursuing the implementation of highly restrictive protections on polar bears at the expense of Inuit, despite the fact that the species is one of the most managed species in the world and is nowhere near being threatened with extinction. This is a direct attack on our rights, culture, hunting practices, conservation and management agreements, and local economies as an indigenous peoples of the Arctic.[12]

The "restrictive protections" Simon refers to were not designed to stop or restrict subsistence hunting, or even the domestic commercial hunting of the polar bear. Simon's remarks were in response to a proposal to have the bear uplisted to Appendix II on the Convention on International Trade in Endangered Species of Wild Fauna and Flora (CITES). Such a listing would result in a ban on the international trade of polar bear parts, and thus would result in restricting the commercial use of the bear to domestic markets. For Simon, what is at stake is a level of cultural autonomy that is itself dependent on using the bear as an intermediary object for economic development. As the proposal was not directed at stopping the local hunting of the bear, Simon's words seem to imply that the ability to maintain one's existence as "indigenous peoples of the Arctic" is dependent on using the bear as a source for economic profit. One could claim that the bear's importance for Inuit identity remains strong because exercising economic self-determination is a necessary condition for cultural autonomy (and thus retaining access to the international market of trophy hunters is vital for "being indigenous"). However, unlike the first example of human-animal relations employed by the hunter, Simon's version of the bear seems to matter culturally on account of what it provides *for* the culture as opposed to what it does *in* it. Here, value and meaning of encountering the bear derives from the economic security it may provide, that is, following its transformation into an international commodity.

Being able to "be human" or express one's unique identity has long found performative space through the way in which we interact with animals. For example, killing animals has been a source to which humans seek to express heterosexual and gender normativity.[13] Particularly following World War II,[14] big-game hunting developed as an activity for the expression of masculinity.[15] Yet, as instrumentalizing animal life has not been reserved to only one culture, neither has it been reserved to one gender. Recent studies have pointed out how women have begun to use the killing of wildlife as a place to assert a degree of femininity into the act of "appreciating nature."[16] The use of animals as passive objects is not just restricted to activities where we strive to kill them. In her discussion of the work of Noske, Birke, and Haraway, Anne Scott describes nonhuman animals as "objects-to-think-with."[17] Maintaining animals as living objects thus provides a similar platform to experience "being human." In *Watching Wildlife*, Cynthia Chris suggests that "we look not only at animals to learn about them, but we also look through animals for ourselves."[18] Similarly, the steady development of ecotourism has allowed a great number of people the chance to capture the natural nonhuman subject on film. In Churchill, Manitoba, where polar bear tourism is a central component to the local economy, one can easily access one of the world's biggest predators via specially designed trucks called Polar Bear Rovers.[19] The draw is to experience polar bears not in zoos but in the wild. Passive objects such as the polar bear could be perceived as providing meaning to the landscape with which they are associated. However, as is explored later in this chapter, it is not just that the bear augments one's experience of the North, but also it is the pristine backdrop of the Arctic that enhances one's experience of the bear. In regards to the relationship between remoteness, rarity, and the pursuit of animal objects, Chris relays Berger's thoughts:

> Zoos, realistic animal toys and the widespread commercial diffusion of animal imagery, all began as animals started to be withdrawn from daily life. One could suppose that such innovations were compensatory. Yet in reality the innovations themselves belonged to the same remorseless movement as was dispersing the animals. The zoos, with their theatrical decor for display, were in fact demonstrations of how animals had been rendered absolutely marginal. The realistic toys increased the demand for the new animal puppet: the urban pet. The reproduction of animals in images—as their biological reproduction in birth becomes a rarer and rarer sight—was competitively forced to make animals ever more exotic and remote.[20]

Chris then goes on to argue that "animals, receding from daily contact with increasingly industrialized and urbanized societies in the late eighteenth through the nineteenth centuries, returned in new commodity forms."[21] Therefore, the more natural or rare a nonhuman object is perceived to be, the higher affective value it will have for the "adventuring" consumer. Also, by thinking of the North as a rugged and pristine landscape, one's experience of being with the bear is augmented by viewing the object through the window of a Polar Bear Rover as opposed to a zoo enclosure. Similar to hunting the bear as a trophy, the value of being able to "shoot" it with a camera or view it up close comes from thinking of the bear as a passive object. Here, the *being with* is an experience enjoyed by the human but is only made possible by disregarding the subjectivity of the bear. That is to say, the bear receives no credit for the encounter; it does not choose to experience the human but rather it is the *skill* of the sportsman/woman, the *perseverance* of the adventure, or the tourist's *appreciation* for wildlife that allows the encounter to occur. So it is the encounter that provides a space for these human qualities to exist. The skill of the hunter, the perseverance of the adventurer, and the appreciation of the tourist would all be greatly diminished if the encounter occurred while the bear paced back and forth in a zoo enclosure. Building off of Berger's reflections, we can begin to understand how it is the degree of rarity of *the encounter of* an animal that helps determines its value. As an individual cow goes through a transformation process in the minds of those who consume her, from cow to beef—a product that can be graded for quality—so too does the polar bear go through a similar process from individual to object. Its grade will be at its purest form when the encounter we have with it produces the concentrated experience. Like a stamp of *organic* or *free-range*, a stamp of the *wild* on the polar bear denotes a higher level of "purity" of the nonhumanness that we crave. Under these conditions, the experiences we obtain from such an encounter serve up a most potent provision for exaltation of one's human (or rather nonanimal) identity. As we will see in the next section, these types of transformations, from subject to object, are also employed when fostering the kinds of affective bonds used to mobilize us for the saving of the nonhuman.

Through the term *existence value* economists and policy makers often operationalize the altruistic belief that species hold some intrinsic value of their own. The Canadian Privy Council explains existence value as "a concept used to refer to the intrinsic value of some asset, normally natural/environmental. It is the value of the benefits derived from the asset's existence alone. For example, a tree can be valued in a number of ways, including its use value (as lumber), an existence value (simply being there), and an option value (value of things that it could be used for). Existence value is separate from the value accruing from any use or potential use of the asset."[22] Why not just use the term *intrinsic value*? Something might have intrinsic value regardless of whether we ascertain its existence. Existence value, on the other hand, represents an attempt to quantify the value of our experience of a thing's existence. That is to say that a particular thing's existence value can only come to be after we are able to encounter it or know of its existence, to grasp it, and thus to some extent appropriate its meaning in accounting for our existence as a human in a nonhuman world.

Species advocacy is often directed at certain species as if some animals, such as the polar bear, have more existence value than others. In 2008, the IUCN reported that 35 percent of birds, 52 percent of amphibians, and 71 percent of reef-building corals have traits that are likely to make them particularly susceptible to a climate-change-driven extinction.[23] Yet it is the polar bear that we have come to know as the poster child of climate change. The bear has been used in everything from leaflets to TV advertisements to giant ice sculptures, all in order to draw attention to the global impacts of climate change. In encouraging the empathetic response to reach for one's pocketbook, advocacy organizations use particular species, which Kalland describes as "flagship species," to establish affinity between donor and the broader political campaign with which the species becomes associated.[24]

The high level of aesthetic or symbolic value that the bear has come to embody in most countries has made it not only a coveted wall mount for some but also an effective tool for garnering support for broader environmental campaigns that are not exclusive to protecting the bear. Using "flagship species" in this manner depends on the existence of a causal link between the plight of an animal and a broader environmental problem. The effectiveness of the "flagship species," however, depends on the degree of affinity that can be established between the potential donor and the animal. As noted by Myers, "diversity is relayed through aesthetic appreciation."[25] This might be a reason why the musk ox or some ice-dwelling extremophile are not "flagship species" used by climate-change campaigners. Advertising campaigns or media blitzes attempt to manufacture narratives of meaning that, as Arran Stibbe points out, do not aim to create affective relationships between people and animals by describing them as individuals "caught up in a life and death battle against the encroachment of humans."[26] Instead, this affective relationship is established through presenting animals as evaluative descriptors of physical attributes, such as "gentle giant," or of the degree of rarity, such as "the world's most endangered big cat."[27] Here, an emotive response can be nurtured by presenting the animal as a congealment of various qualities that are valued as a zoological aesthetic, part of a broader natural, nonhuman heritage. Stibbe's summary of the WWF's most commonly used adjectives to describe its most advertised twelve animals denotes such an attempt to present the task of species conservation in more palatable terms: "largest, larger, plentiful, mighty, huge, powerful, endangered, numerous, widespread, wondrous, remarkable, lovable, charismatic, popular, small, shy, majestic, silver, anadromous."[28]

The call to preserve this part of our natural nonhuman heritage, a part that we often refer to as the "animal kingdom," can serve as particularly ripe grounds for species advocacy. As Turner explains, existence value relates to several anthropocentric dispositions toward the instrumental and noninstrumental value of animals.[29] Species are the places in which the "lovability" and "majesty" of the nonhuman world are stored, and thus, as Turner notes, support for species protection may be brought on by an intergenerational altruism directed at ensuring some of the most majestic and lovable creatures remain available for our future generations.[30] This intergenerational altruism has been recently activated as a result of climate change threatening our children's ability to enjoy our mighty and majestic polar bear. The ability to appeal to this intergenerational altruism has not been lost upon nongovernmental organizations such as the WWF:

> The planet is rapidly moving towards a tipping point with climate change impacts on polar bears. Unless immediate action is taken by responsible governments, we may be relegating polar bears to extinction in the wild within the lifetime of our children.[31]

By being thought of as natural nonhuman objects, animals are able to serve as points of reference for our humanity. However, by relating to species as passive objects, without subjectivities independent from our own, we leave ourselves open to becoming emotionally attached to experiential tropes. This has direct implications for how the value and definition of the well-being of any given species is defined or brought in and out of moral existence. While referring to Burt in "Developing Animals," Brower[32] makes a similar point:

> As Burt notes, "The idea that the animal as a natural nonhuman object is always automatically corrupted or falsified as soon as it is visually troped" denies the possibility of any appropriate human-animal relation. (188)

Advocacy depends on retaining a level of affective commitment from us. However, it becomes clear that this affective commitment is fostered toward our representations of species, not toward the individuals that make up the species.

Conclusion

This chapter explores the ways in which animals are made into passive objects. This is not a facet intrinsic to all human-nonhuman relations but is, as argued here, an aspect of a particular humancentric model. By ignoring the subjectivity of animals, we are provided a performative space in which we can express lived experiences exclusive to *being human* in a *nonhuman world.* Examining the specific political consequences of this is outside the scope of this chapter.[33] However, exploring this model of animal-human relations shows how affinity between the human subject and animal object can be articulated equally through killing and saving. Both acts are guided by a moral imperative to retain our encounters with animals as spaces for the production of a distinction. This distinction, while allowing us a place to *be* human, also, in consequence, reduces the space for the nonhuman to be valued or, in a sense, for the nonhuman to exist outside our instrumental use of them.

Endnotes

1. Jeremy J. Schmidt and Martha Dowsley, "Hunting with Polar Bears: Problems with the Passive Properties of the Commons," *Human Ecology* 38, no. 3 (April 15, 2010): 383.

2. Ibid.

3. André Krebber, "Anthropocentrism and Reason in Dialectic of Enlightenment: Environmental Crisis and Animal Subject," in *Anthropocentrism: Humans, Animals, Environments*, ed. Rob Boddice (Leiden and Boston: Brill, 2011).

4. See Robyn Eckersley, *The Green State: Rethinking Democracy and Sovereignty* (Cambridge, MA: MIT Press, 2004), and Andrew Dobson, *Green Political Thought*, 3rd ed. (London: Routledge, 2000); John S. Dryzek, *Rational Ecology: Environment and Political Economy* (Oxford: Basil Blackwell, 1987); John S. Dryzek, notes to pages 7–9 and notes to pages 9–20 in *Discursive Democracy: Politics, Policy and Political Science* (Cambridge: Cambridge University Press, 1990); Tim Hayward, *Ecological Thought: An Introduction* (Cambridge, MA: Polity Press, 1994); Tim Hayward, *Political Theory and Ecological Values* (New York: St. Martin's Press, 1998); Mary Mellor, *Feminism and Ecology* (New York: New York University Press, 1997); Douglas Torgerson, *The Promise of Green Politics: Environmentalism and the Public Sphere* (Durham: Duke University Press, 1999); Val Plumwood, *Feminism and the Mastery of Nature* (London:

Routledge, 1993); and Val Plumwood, *Environmental Culture: The Ecological Crisis of Reason* (London: Routledge, 2002).

5. Dario Martinelli, "Anthropocentrism as a Social Phenomenon: Semiotic and Ethical Implications," *Social Semiotics* 18, no. 1 (2008): 79–99.

6. Robert Garner, "Wildlife Conservation and the Moral Status of Animals," *Environmental Politics* 3, no. 1 (1994): 114–129.

7. David Klein, "Management and Conservation of Wildlife in a Changing Arctic Environment," in *Arctic Climate Impact Assessment* (Cambridge: Cambridge University Press: 2005).

8. Jane George, "Polar Bear Management in Nunavut: A Conservation Tight-Rope," *Nunatsiaq News*, accessed April 18, 2014, http://www.nunatsiaqonline.ca/archives/nunavut991130/nvt91119_07.html.

9. Milton M. R. Freeman and George W. Wenzel, "The Nature and Significance of Polar Bear Conservation Hunting in the Canadian Arctic," *Arctic* 59, no. 1 (2006).

10. See Plumwood, *Environmental Culture* and *Feminism and the Mastery of Nature*.

11. For what Eckersley describes as a "critical political ecology," see Eckersley, *The Green State*, p. 9. However, for a discussion on the animal and critical theory specifically, see Marco Maurizi, "The Dialectical Animal: Nature and Philosophy of History in Adorno, Horkheimer and Marcuse," *Journal for Critical Animal Studies* 10 (2012): 1.

12. Inuit Tapiriit Kanatami, *Inuit Disappointed at US Activism on CITES Polar Bear Up-Listing Proposal*, accessed April 18, 2014, https://www.itk.ca/media/media-release/inuit-disappointed-us-activism-cites-polar-bear-listing-proposal.

13. See Donna Haraway, "The Promises of Monsters: A Regenerative Politics for Inappropriate/D Others," *Cultural Studies* (1992): 295–337, and Harriet Ritvo, *The Animal Estate: The English and Other Creatures in the Victorian Age* (Cambridge, MA: Harvard University Press, 1987).

14. Andrea L. Smalley, "'I Just Like to Kill Things': Women, Men and the Gender of Sport Hunting in the United States, 1940–1973," *Gender & History* 17, no. 1 (2005): 184.

15. For studies that connect hunting and masculinity, see Lisa M. Fine, "Rites of Man, Rites of Passage: Hunting and Masculinity at REO Motors of Lansing, Michigan, 1945–1975," *Journal of Social History* 33 (2000): 805–23; E. Anthony Rotundo, *American Manhood: Transformations in Masculinity from the Revolution to the Modern Era* (New York: Basic Books, 1993); Elliott Gorn, *The Manly Art* (Ithaca, NY: Cornell University Press, 1986); Gail Bederman, *Manliness and Civilization: A Cultural History of Gender and Race in the United States, 1880–1917* (Chicago: University of Chicago Press, 1995); Michael Kimmel, *Manhood in America: A Cultural History* (New York: The Free Press, 1996).

16. Smalley, "'I Just Like to Kill Things.'"

17. Anne Scott, "Trafficking in Monstrosity: Conceptualizations of 'Nature' within Feminist Cyborg Discourses," *Feminist Theory* 2 (December 2001): 371, cited by Chilla Bulbeck, *Facing the Wild: Ecotourism, Conservation and Animal Encounters* (London: Taylor and Francis, 2012).

18. Cynthia Chris, *Watching Wildlife* (Minneapolis: University of Minnesota Press, 2006).

19. For an overview of the polar bear watching industry in Churchill, Manitoba, see R. Harvey Lemelin, "The Gawk, the Glance, and the Gaze: Ocular Consumption and Polar Bear Tourism in Churchill, Manitoba, Canada," *Current Issues in Tourism* 9, no. 6 (2006): 516–534.

20. John Berger, *About Looking* (New York: Vintage, 2011), quoted in Chris, *Watching Wildlife*, x.

21. Ibid.

22. Government of Canada, Privy Council Office, "Existence Value," *Glossary*. Last modified May 6, 2006, accessed April 18, 2014, http://www.pco-bcp.gc.ca/raoics-srdc/default.asp?Language=E&Page=glossary.

23. Wendy Foden et al., "Species Susceptibility to Climate Change Impacts," in *Wildlife in a Changing World: An Analysis of the 2008 IUCN Red List of Threatened Species* (Gland, Switzerland: IUCN, 2009), 77.

24. Arne Kalland, "Whose Whale Is That? Diverting the Commodity Path," in *Elephants and Whales: Resources for Whom*, eds. Milton Freeman and U. P. Kreuter (New York: Taylor & Francis, 1994).

25. Norman Myers, *The Sinking Ark: A New Look at the Problem of Disappearing Species* (New York: Pergamon Press, 1979).

26. Arran Stibbe, *Animals Erased: Discourse, Ecology, and Reconnection with the Natural World* (Middletown, CT: Wesleyan University Press, 2012), 74.

27. Ibid.

28. Arran Stibbe, "Counter-Discourses and the Relationship Between Humans and Other Animals," *Anthrozoos: A Multidisciplinary Journal of the Interactions of People & Animals* 18, no. 1 (2005): 10.

29. R. Kerry Turner et al., "Valuing Nature: Lessons Learned and Future Research Directions," *Ecological Economics* 46, no. 3 (2003): 497–499.

30. Ibid.

31. WWF (World Wildlife Fund), "A Push for Change for Polar Bears," *World Wildlife Fund* e-newsletter (2009), 1–3.

32. Matthew Brower, *Developing Animals: Wildlife and Early American Photography* (Minneapolis: University of Minnesota Press, 2011), xvi, citing Jonathan Burt, *Animals in Film* (London: Reaktion, 2002), 188.

33. See Russell Fielding, "Indigeneity and Ecology in Iñupiaq and Faroese Whaling" (this volume), for a discussion on the politics behind the "save the whales" campaigns and the consequences this had for conservation.

Bibliography

Berger, John. *About Looking*, New York: Vintage, 2011.

Brower, Matthew. *Developing Animals: Wildlife and Early American Photography*. Minneapolis: University of Minnesota Press, 2011.

Bulbeck, Chilla. *Facing the Wild: Ecotourism, Conservation and Animal Encounters*. London: Taylor and Francis, 2012.

Burt, Jonathan. *Animals in Film*. London: Reaktion, 2002.

Chris, Cynthia. *Watching Wildlife*. Minneapolis: University of Minnesota Press, 2006.

Eckersley, Robyn. *The Green State: Rethinking Democracy and Sovereignty*. Cambridge, MA: MIT Press, 2004.

Foden, Wendy, Georgina M. Mace, Jean-Christophe Vié, Ariadne Angulo, Stuart H. M. Butchart, Lyndon DeVantier, Holly T. Dublin, Alexander Gutsche, Simon Stuart, and Emre Turak. "Species Susceptibility to Climate Change Impacts." In *Wildlife in a Changing World: An Analysis of the 2008 IUCN Red List Of Threatened Species*. Gland, Sweden: IUCN, 2009.

Freeman, Milton M. R., and George W. Wenzel. "The Nature and Significance of Polar Bear Conservation Hunting in the Canadian Arctic." *Arctic* 59, no. 1 (2006).

Garner, Robert. "Wildlife Conservation and the Moral Status of Animals." *Environmental Politics* 3, no. 1 (1994): 114–129.

George, Jane. "Polar Bear Management in Nunavut: A Conservation Tight-Rope." *Nunatsiaq News*, November 19, 1999. Accessed April 18, 2014, http://www.nunatsiaqonline.ca/archives/nunavut991130/nvt91119_07.html.

Government of Canada, Privy Council Office. "Existence Value." Last modified May 6, 2006. Accessed April 18, 2014, http://www.pco-bcp.gc.ca/raoics-srdc/default.asp?Language=E&Page=glossary.

Haraway, Donna. "The Promises of Monsters: A Regenerative Politics for Inappropriate/D Others." In *Cultural Studies*, edited by Larry Grossberg, Cary Nelson, and Paula Treichler. New York: Routledge, 1992.

Inuit Tapiriit Kanatami. "Inuit Disappointed at US Activism on CITES Polar Bear UP-Listing Proposal." Accessed April 18, 2014, https://www.itk.ca/media/media-release/inuit-disappointed-us-activism-cites-polar-bear-listing-proposal.

Kalland, Arne. "Whose Whale Is That? Diverting the Commodity Path." In *Elephants and Whales: Resources for Whom*, edited by Milton Freeman and U. P. Kreuter. New York: Taylor & Francis, 1994.

Klein, David. "Management and Conservation of Wildlife in a Changing Arctic Environment." In *Arctic Climate Impact Assessment.* Cambridge: Cambridge University Press, 2005.

Krebber, André. "Anthropocentrism and Reason in Dialectic of Enlightenment: Environmental Crisis and Animal Subject." In *Anthropocentrism: Humans, Animals, Environments,* edited by Rob Boddice. Leiden and Boston: Brill, 2011.

Lemelin, R. Harvey. "The Gawk, the Glance, and the Gaze: Ocular Consumption and Polar Bear Tourism in Churchill, Manitoba, Canada." *Current Issues in Tourism* 9, no. 6 (2006): 516–534.

Martinelli, Dario. "Anthropocentrism as a Social Phenomenon: Semiotic and Ethical Implications." *Social Semiotics* 18, no. 1 (2008): 79–99.

Maurizi, Marco. "The Dialectical Animal: Nature and Philosophy of History in Adorno, Horkheimer and Marcuse." *Journal for Critical Animal Studies* 10 (2012).

Myers, Norman. *The Sinking Ark: A New Look at the Problem of Disappearing Species.* New York: Pergamon Press. 1979.

Plumwood, Val. *Environmental Culture: The Ecological Crisis of Reason,* volume 3 of *Environmental Philosophies.* New York: Routledge, 2005.

———. *Feminism and the Mastery of Nature.* New York: Routledge, 2002.

Ritvo, Harriet. *The Animal Estate: The English and Other Creatures in the Victorian Age.* Cambridge, MA: Harvard University Press, 1987.

Smalley, Andrea L. "'I Just Like to Kill Things': Women, Men and the Gender of Sport Hunting in the United States, 1940–1973." *Gender & History* 17, no. 1 (2005): 183–209.

Stibbe, Arran. *Animals Erased: Discourse, Ecology, and Reconnection with the Natural World.* Middletown, CT: Wesleyan University Press, 2012.

———. "Counter-Discourses and the Relationship Between Humans and Other Animals." *Anthrozoos: A Multidisciplinary Journal of the Interactions of People & Animals* 18, no. 1 (2005): 3–17.

Schmidt, Jeremy J., and Martha Dowsley. "Hunting with Polar Bears: Problems with the Passive Properties of the Commons." *Human Ecology* 38, no. 3 (April 15, 2010): 377–87. doi:10.1007/s10745-010-9328-0.

Turner, R. Kerry, Jouni Paavola, Philip Cooper, Stephen Farber, Valma Jessamy, and Stavros Georgiou. "Valuing Nature: Lessons Learned and Future Research Directions." *Ecological Economics* 46, no. 3 (2003): 493–510.

WWF. "A Push for Change for Polar Bears." *World Wildlife Fund E-Newsletter.* March 6, 2009. Accessed April 18, 2014. http://wwf.panda.org/what_we_do/where_we_work/arctic/?158181/wwf-push-for-change-for-polar-bears-In-2009.

6

Bare Life and Bear Love

Masculinity, Capital, and Arctic Animals in the Nineteenth-Century North

john miller / UNIVERSITY OF SHEFFIELD

THE CAREER OF HORATIO NELSON, THAT MOST CELEBRATED OF British heroes, was famously almost ended before it began by an encounter with a polar bear. As a fourteen-year-old midshipman on board *HMS Carcass*, engaged in a "voyage of discovery towards the North Pole,"[1] Nelson and his shipmates found themselves stuck fast in ice near Spitzbergen. Seeing a bear, Nelson decided to present its skin to his father and set off in pursuit with a companion, armed only with a "rusty musket."[2] The gun misfired at the vital moment, leaving Nelson at the bear's mercy. According to Robert Southey, the young hero approached the situation with significant pluck: "'Never mind,' exclaimed Horatio, 'do but let me get a blow at this devil with the butt-end of my musket, and we shall have him.'"[3] The captain of the *Carcass*, Skeffington Lutwidge, observed the events from the deck and fired the ship's cannon, startling the bear, which was then separated from the imperiled teenager by a split in the ice.

What the enraged Lutwidge saw as an act of adolescent folly has nonetheless become an integral part of Nelson's mythology. Richard Westall, a noted painter of historical subjects, was commissioned in 1806, the year after Nelson's death, to produce an illustration for John McArthur and James Stannier Clarke's *Life of Admiral Lord Nelson, KB*, a text designed to

Fig. 6.1. *Nelson and the Bear* by Richard Westall, 1806.

be "held out as an example of Heroism and professional Talent to future generations."[4] Westall's *Nelson and the Bear* (fig. 6.1) is a spectacle of youthful virility that anticipates more momentous feats to come. For another of Nelson's biographers, Captain A. T. Mahan, Nelson was the "first of British seamen, whom the gifts of Nature and course of History have united to make, in his victories and in their results, the representative figure of the greatest sea-power that the world has known." In the narrative of Nelson's involvement in Britain's ascent to global preeminence, the arctic incident becomes, in Mahan's words, "hackneyed by frequent repetition."[5] The bear functions as a text in which Britain's future is foretold. Nelson's intended present of its hide to his father suggests genealogy as well as prolepsis. The vanquished beast is gauged to provide a guarantee of Nelson's maturation into an ideal of British manhood in consummation of his father's aspiration. More than just a skin, the thwarted gift is a promise of extrovert masculine accomplishment performed on the global stage.

The story of Nelson's bear is a useful example of the dominant paradigm of British representations of the Arctic in the early nineteenth century. Representations of the Far North routinely imagined it as a space of existential desolation, a terrifyingly isolated realm with dark, supernatural overtones. As Chauncey Loomis summarizes, "Arctic Nature was somehow vaster, more mysterious and more terrible than elsewhere on the globe."[6] Nelson's intention to "get a blow at this *devil* with the butt-end of [his] musket" (emphasis added) inserts the bear into the literary tradition of the fiendish North and conveys an important wider implication of this patriotic display. The imposition of British masculine energies on the Arctic is also the assertion of the human over the monster; nationalist discourses coincide with discourses of species. Westall's picture hinges on an aesthetics of domination, with Nelson's foolhardy endeavor elevated to the larger contest of British civility against the wild North.

The Arctic consequently serves as a privileged arena in the formation of Britain's imperialist geography in the nineteenth century: the bleaker the environment, the more resounding the sense of national triumph. British representations of the North, as Loomis concludes, "took on an almost allegorical significance."[7] In slightly different terms, the construction of the nineteenth-century Arctic can be read as a repertoire of metonyms. Nelson's bear is a sign of its inhospitable habitat; Nelson, in turn, encapsulates Britain's naval supremacy as its "representative figure" to return to Mahan's words. Britain's arctic expeditions are recurrently the stuff of romance; bodies, both human and animal, are assimilated into larger, more abstract narratives of patriotic masculinity. A good deal of scholarship has articulated the evident disassociation of the material North from its representation in the context of British arctic endeavors. Sarah Moss argues that "exploration is . . . largely about incorporating strangeness into nationhood, returning to tell the tale that then becomes part of the home culture."[8] The famed whiteness of the hyperborean world routinely suggested a blank and empty space, which, as Jen Hill argues, erased "the area's geographical complexity."[9] In these largely homogeneous depictions, the pervasive representation of arctic animals is for the most part reducible to two modes of signification; as Robert G. David explains, "any consideration of Arctic fauna" usually returned to the themes of "danger and commerce."[10] Contacts with arctic animals are routinely inscribed within interconnected discourses of the heroic and the economic. Around two hundred years before the current scramble for arctic resources, the Far North had already begun its entry into global capital in a manner that had as much to do with image and the imaginary as it had to do with material utility.

In the decades following Nelson's adventures on the *Carcass*, British arctic interests intensified, motivated partly by the question of what to do with the nation's sailors, idling on half pay after the end of the Napoleonic Wars. The ongoing search for the pole and the Northwest Passage appeared as emblems of a burgeoning national confidence and as a powerful index of Britain's international standing that was also nourished by the lucrative harvests of the whaling fleet. While Charles Dickens felt himself "filled with a sort of sacred joy"[11] at the prospect of an expedition to the Canadian Arctic in the 1840s, a poem in the arctic adventurer Sir William Parry's *North Georgia Gazette* from 1821 strikes an altogether different note:

> Fired with fresh ardour and with bold intent,
> Our minds shall, like our prows be westward bent,
> Until Pacific's waves pour forth sweet sounds,
> Chiming to us like—Twenty thousand pounds![12]

Evidently, the immediate value of the discovery of the Northwest Passage to Britain's navy was the prize money offered by the Admiralty in anticipation of the mercantile benefits a northern sea route to the west might open up.[13] In contrast, as the century progressed, some aspects of arctic exploration would become associated with a certain purity of motive. There was very little by way of an economic or strategic rationale in the quest for the North Pole, for example; for Russell Potter, the pole's "uselessness, except as a symbol, was given from the start."[14] The *North Georgia Gazette*'s counterpoint between the lofty tone of the sailors' "fresh ardour" and "bold intent" and the blunt calculation of the Admiralty's £20,000, however, exemplifies the entanglement of nationalist identity politics with economic discourses. Even expeditions that seemed bent only on glory were still intimately connected with a commercial agenda in two significant ways.

First, voyages of scientific exploration, despite their apparent distance from mercantile enterprises, were involved with the whaling fleet's exploitation of arctic sea mammals, not least through the exchange of mariners between naval and whaling missions. To give one prominent example among many, John Barrow, the Admiralty Secretary and a significant driving force behind polar exploration, had cut his teeth on a youthful whaling voyage.[15] Second, any supposition of the financial innocence of British arctic ambitions neglects the mutual involvement of the material and the semiotic in a century notable for the rise of what Thomas Richards calls the "cultural forms of consumerism."[16] Britain's increasing interest in the Arctic coincided, as Potter reminds us, with "the dawn of a new visual culture of mass history, mass images, and a mass reading public."[17] Arctic missions drew extended coverage in emerging illustrated newspapers with the superabundance of travel narratives testifying to a deep appetite for heroic tales of exploration. The *Illustrated London News* was one of the publications at the forefront of the marketing of the iconography of the North. The 1849 *Pictures of the Polar Regions* (fig. 6.2) provides a catalogue of arctic stereotypes that compresses the North into a set of stock images: the glittering berg, the breaching whale, the lumbering walruses, and above them the aurora borealis framing the intrepid vessels. In addition to the illustrated newspapers, there were, Potter recounts, "no fewer than 60 Arctic shows" in Britain between 1818 and 1883, including exhibitions and panoramas that recycled the same version of the Arctic.[18] The proliferation of emerging visual technologies, "these 'new media' of the Victorian era" as Potter calls them,[19] provided significant financial opportunities for expeditions that were otherwise of limited direct economic value. Capital and aesthetics converge and, as the *Illustrated London News* reveals, animals are central to this as visual commodities around which narratives of arctic heroism are recurrently structured.

Heroism in this context is part of the spectacle of imperial capital, or, as Francis Spufford argues, there came to be a "close relationship between glory and booty."[20] Evidently, the sense of the nineteenth-century North

Fig. 6.2. "Pictures of the Polar Regions," *Illustrated London News*, 1849.

as an *imaginary* zone in British colonial discourse positions the period's arctic narratives at an uncomfortable remove from ecocritical or animal studies perspectives, committed to foregrounding material environments and lived histories, both human and nonhuman. Crucially, however, rather than amounting to a consistent expression of nationalist masculinities, the conjunction of heroism and capital in nineteenth-century British representations of arctic animals is already internally contested. It is with a less commonly explored, counterhegemonic sense of the Arctic's radical ambiguity that this chapter is concerned.

James Hogg's novella "The Surpassing Adventures of Allan Gordon," published posthumously in 1837, is perhaps the period's oddest tale of British arctic endeavors and one that provides a startling counterpoint to mainstream representations of arctic animals in relation to economic discourses and ideals of masculine self-assertion. Contrary to what had become established as a conventional pattern, Hogg's eponymous hero discovers a surprising intimacy with an arctic creature that contests the dominant dichotomy of the heroic adventurer and the savage North and reveals an alternative experience of the discourse of capital. "The Surpassing Adventures of Allan Gordon," therefore, sounds a distinctive note in its depiction of the intertwining questions of nation, capital, and aesthetics in meetings with arctic animals. The North, as Sarah Jaquette Ray and Kevin Maier contend in the introduction to this volume, is "less a place than a set of cultural values and associations." Crucially, "Allan Gordon" expands understandings of the range of cultural factors at work in the social construction of the Arctic in nineteenth-century Britain, investigating the region not from the empowered perspective favored in the analysis of Spufford, Loomis, and others but through the sensibility of a tormented social outsider.

Hogg is known mainly as the author of the 1824 novel *The Private Memoirs and Confessions of a Justified Sinner* and also as the Ettrick Shepherd, the moniker under which he came to be a notable contributor to Romantic literary culture. As a Scot with a deep investment in the rural working class, Hogg wrote from outside the nineteenth-century centers of power and offers a different angle on Britain's arctic mythology. As Penny Fielding writes, "'Allan Gordon' plunges its readers into a comically grotesque vision of the failure of power, of the self, of sexuality and of language."[21] While Spufford's claim that in "the wilderness of the ice, men were shown as moral pin-ups"[22] rings true for the majority of arctic narratives, Hogg's tale is interesting precisely for its contestation of such normative formulations. More specifically, it is Hogg's portrayal of the eponymous hero's relationship with a polar bear that makes this a useful text for an environmental humanities reflection on the cultural production of the Far North in the nineteenth century. If, as Cian Duffy writes, in confirmation of most scholarship on the nineteenth-century British Arctic to date, "the discovery of the true polar sublime . . . was . . . the discovery of the *inhuman*" (original emphasis),[23] Hogg's story is significant for the games it plays with the distinction of human and inhuman, a topic that is implicit but undeveloped in Fielding's important essay on "Allan Gordon."

In short, "The Surpassing Adventures of Allan Gordon" is the story of a shipwrecked whaler's friendship with an orphaned bear cub he adopts and names Nancy after a childhood sweetheart. The narrative begins with Gordon, an apprentice tailor from Aberdeenshire, quarreling with his employer and going to sea aboard a whaling ship, which becomes wrecked on an iceberg. Gordon is the only survivor and is destined to a lonely existence until, after killing an intruding bear, he discovers her newborn cub. At first the bear is entirely dependent on the sailor, but as the narrative progresses Nancy takes on the nurturing role, feeding and protecting her human companion. Problems appear in Gordon's relationship with Nancy when the iceberg on which the ship is stranded drifts into a Norwegian colony in Greenland. Gordon joins the residents, but when he determines to find a wife he learns that Nancy, who has been used to sharing his bed, refuses to "suffer another to lye in [his] arms in her presence."[24]

Once Nancy has hibernated for the winter, Gordon is free to marry (and also to take a number of mistresses), although when the bear wakes he again faces a problem: "I could not leave my wife and my other supposed beautiful inamorata to sleep with a huge white bear" (37). The upshot is that Nancy leaves, and Gordon's life becomes a "mere blank in existence" (38). Finally, a horde of white bears attacks the colony and the terrified population barricade themselves in a cave to suffocate in preference to being torn apart. Gordon leads a counterattack only to realize that the bears have outflanked him with disastrous consequences: "The women were all seized first particularly the young ones prostrated and devoured." The colony is decimated, but Gordon is saved by Nancy, who, as he describes in the awkward syntax characteristic of Hogg's story, "whimpered her joy groveled on the snow and licked my feet my knees my hands" (40). He consequently abandons the few surviving colonists and heads off across Greenland with Nancy, who delivers him to a cache of canoes before returning to the ravening bears. Gordon is rescued by a passing Dutch vessel.

Despite the singularity of the plot, the popular discourse of the nineteenth-century Arctic is clearly apparent in Hogg's mise-en-scène. A bear arrives on the ship first, conventionally, as a "foraging monster" and then as an "apparition" (10). As Sarah Moss argues, the "bears clearly personify the physical and epistemological threat posed by the Arctic."[25] Gordon duly dispatches the eldritch intruder with "two deadly stabs below the fifth rib toward the heart" (10) in an entirely unsurprising arctic scenario of the voyager mastering the denizens of this uncanny domain. Hogg's hero is no Lord Nelson, however. Rather, as he frequently reminds us, Gordon is a man of a "cowardly and timorous" disposition (10) who cuts a forlorn figure in the normative cast of the British Arctic. Confronted by the bear, he confesses to a "palpitating heart" and experiences what in the context would appear a quite unusual pang of regret on killing Nancy's mother. His distance from the expected traits of the arctic hero is confirmed by his immoderate attention to the ship's grog as he initially drinks heavily and cuts a wretched figure, drunk on the "grand elixir" (7) as he calls it.

Gordon's is very much a limit-experience through which Hogg gradually brings the nationalist tropes to the fore in representations of early nineteenth-century arctic exploration. If Nelson's intended gift of the bear skin to his father forges a connection between colonial power and paternity, Hogg's story, by contrast, is far more interested, as Fielding has shown, in mothers and motherhood. More specifically, to a large extent "Allan Gordon" is a tale of absent mothers. The first instance of this motif comes with Gordon's poignant description of the bear cub's response to her mother's death:

> When it found its parent's skin it uttered a bleat of joy and the tears actually streamed from its eyes. It went on uttering the same sort of joyful sounds that a foal does which has been long kept from its dam. It went round and round and licked the skin for very fondness but alas it was always looking for what it could not get, the mother's exhausted dug. . . .
>
> It showed neither hostility to nor fear of me its mother was its home, its guide and director, and it had no other. So after going several times round with the most hopeless looks that I ever witnessed it laid itself down in a round form to die contented but its groans were not to be borne. I proffered it some biscuit . . . It took the first small piece shily and timorously but the rest it took so voraciously that it scarcely took any time to masticate it and I was afraid it would choak. It had been clearly on the very point of famishing. (11)

Gordon's account of the cub's grief contains some egregious moments of anthropomorphism, most notably the tears streaming from Nancy's eyes. Nonetheless, this encounter with Nancy emphasizes a deep sense of the animal's affective and affecting experience. Far from a monstrous hide on which to inscribe the story of Britain's global empowerment, the mother's body here is a site only of disappointed longing as the cub "licks the skin for very fondness." Given the prominence of taxidermy in nineteenth-century Britain, there is a profound resonance to Hogg's attention to skin in connection to animal affect rather than colonial display. The intimacy this moment initiates is emphasized by a series of textual echoes. Gordon's timorous nature finds a counterpart in the bear "shily and timorously" taking a biscuit from his hand. Most significantly, the bear's mournful investigation for her "mother's exhausted dug" is oddly paralleled by Gordon's own experience on the iceberg.

While the tailor initially functions as something like the cub's surrogate mother, he is also, in turn, himself a creature in search of a mother, again an aspect of the text that provides a central focus of Fielding's analysis. Desperate for fresh water, he leaves the ship for the berg it is wrecked on and lies on the ice, where he "sucked and sucked till [he] could hold no more" (5). Not long after this, he finds the ship's grog and uses a set of bellows to suck up the "nectar" with "merry and liberal sucks" (7). Sucking in these early stages provides immediate consolation and highlights the theme of maternity. After his first binge, Gordon remembers that he had "actually lain in a trance for at least a month in that closet in utter darkness the door having been closed no doubt by myself but how or when I knew not" (6). There is a hint of ursine hibernation in this moment that anticipates Gordon's later entry into the world of bears, but it is also very obviously a retirement into an amniotic space that closes him into relative safety, with the umbilicus of the bellows tubing linking him to the motherly comfort of grog.

There is more orality at the novella's denouement as the attacking bears "ravaged all the women and then *sucked their blood* from the throats" (38). Everyone is sucking, or trying to find something to suck, in Hogg's narrative. Fielding draws particularly on Julia Kristeva's work to explain how "the limitless arctic becomes a gigantic maternal body."[26] For Fielding,

Hogg offers an example of what Barbara Creed has called the "Monstrous-Feminine," an experience that pivots on a horror toward "mothering" and female reproductive functions.[27] This trope forms another part of the novella's debt to the conventional motifs of the nineteenth-century Arctic. The Arctic once more functions as an imaginary zone, the North as the great blank mystery of the unconscious in which, in this instance, the narrator discovers an intense sexual anxiety. Gordon's weird erotic adventures in the Norwegian colony further illustrate his dismay at the female body. On first arriving he styles himself as something of a Lothario, intending to take at least three wives. Being restricted to one, he begins an affair, he supposes, with the beautiful Zeluki, only to discover that she remained a "virgin blooming" (37) while all the "ugly" women of the village crept into his chamber to trick him into impregnating them so that they become "the ill-favoured disgusting mothers" (38) of his illegitimate offspring. Fielding, for very good reasons, finds it difficult to "easily read ["Allan Gordon"] as a feminist text,"[28] although it does clearly offer an alternative account of masculinity. If the Arctic is conventionally an arena where men have to be men, Hogg's construction of a lustful but disturbed central figure transposes the convergence of manhood and nationhood into an eidetic other-world. The exaggeration of the trope of heroic encounter between adventurer and bear into a full-scale pitched battle between the colony and dozens of bears further contributes to the ambience of hallucinatory excess that undercuts the conventional struggle of adventurer against animal.

In broader terms, the motif of sucking reveals a fascination with drawing what is inside outside, at the same time that the dreamlike atmosphere of Hogg's prose imagines the outside as a kind of inside, the landscape as a psychic projection. Crucially, in dismantling the explorer/hero figure, Hogg's arctic imaginary serves not to extol but to disrupt the category of the human. Particularly, Hogg's novella is a useful contribution to a process described by Matthew Calarco, after Giorgio Agamben, as "jamming the anthropological machine," which is to say short-circuiting "the mechanism underlying our current means of determining the human-animal distinction."[29] To explain further, again in Calarco's words, "human beings who take an essentially animal form are used to mark the constructive outside of humanity proper: the infant savage, the wolf-man, the werewolf, the slave or the barbarian."[30] Gordon's adventures with Nancy work very neatly with this formulation, supported by a geography that functions as "the constructive outside of humanity" par excellence: the last great wilderness, in common parlance, against which the human domain is defined.

The point, however, of Agamben's theorization of the anthropological machine is that "the structure . . . that delimits the contours of the human is perfectly ironic and empty."[31] In *Homo Sacer*, Agamben starts out from Michel Foucault's "examination of the technologies of the self by which processes of subjectivization bring the individual to bind himself to his own identity and consciousness and, at the same time, to an external power."[32] Hogg's novella immerses its hero into a world beyond such "technologies of the self," outside the ship's relative security and into what Agamben might call a "zone of indistinction." Here any sense of species allegiance collapses. Gordon notices, for example, that the arctic seabirds "regarded [him] only as a fellow creature" (17). This kind of indifferentiation between human and nonhuman evokes Agamben's idea of "bare life," "the simple fact of living common to all living beings"—put simply, human animality.[33] "Bare life," for Agamben, is life without recognition as a citizen, excluded from political significance. All through the narrative, Gordon functions in one way or another as an outsider. Most significantly, from his early life as a victimized apprentice tailor, he suffers under the operation of something that approximates to Foucault's "external power." Recounting his life as an apprentice in Aberdeen, he notes of his employer, "I was always obliged to *Sir* and *master* him and if by chance I called him by any other name I got the length of the needle in my flesh instantly and a well threshed skin when we got out of sight" (1). Trapped in his apprenticeship for "fear of the fine that would fall on [his] poor father" (1), Gordon's life is dominated from the start by economic power structures.

Crucially, the image of the needle returns after his escape to sea. As the ship approaches the North Pole, the compass goes haywire; Gordon reports "there was certainly something peculiar in our situation for the *needle had no power not a grain*" (3, emphasis added). The adventures that follow both respond to and evade the needle as the sign of abuse: the Arctic in Gordon's imagination is both an expression of and a retreat from violence. His exile onto the berg materializes the laboring subject's exposure to the traumatic effects of capital. Gordon's bond with the bear, then, is structured around the commonality between them as bodies beyond ethics. Bare life facilitates bear love as Gordon confesses his sincere and intense affection for his companion (13). Evidently, this alliance presents interpretive difficulties. Both Gordon and Nancy are engaged in acts of transference. Nancy, Gordon summarizes, "transferred her love from her mother to me her only protector" (13), while the bear in turn is also a stand-in:

> She answered to her name and came at my bidding and when we walked out upon the ice I took her paw in my arm and learned her to walk upright. A pretty couple certainly we were I dressed like a gentleman in my late captain's holiday clothes and she walking arm in arm with me with her short steps her long taper neck and unfeasible long head.

The obvious point of comparison for "Allan Gordon" in more canonical nineteenth-century writing is Mary Shelley's *Frankenstein*, but there is something of the flavor of Jane Austen to Hogg's description of this performance of conventional heterosexual intimacy in the promenading "pretty couple" of a naval captain and his belle. There is also a hint of Daniel Defoe's *Robinson Crusoe* here. As Moss points out, Gordon's relation to Nancy evokes "Crusoe's relationship with Man Friday and to some extent can be read in a similar way as the triumph of *homo economicus* over the native representative of strange space."[34] The love between the tailor and the bear appears as a paragon of respectability, a representational pattern that marks the normalization of desire by capital.

Except, of course, the bear's "long taper neck and unfeasible long head" render the scene entirely absurd. Hogg's surreal twist adds both pathos and irony to the way that the victim of capital comes to share in its triumphs as the man about town in his holiday clothes. The substitution of a full-grown bear for the wife on his arm emphasizes the distance of the working-class laborer from the locations of power and privilege while exploding bourgeois conventions into a hallucinatory other-world. Although a key part of the political force of Hogg's story is Gordon's deprivation of the human realm, vitally this privation also functions as a condition of ethical possibility in which relationships to the nonhuman can be reconceived. As Moss argues, "What Allan achieves in his relationship with Nancy, and what Hogg achieves in his writing of the Arctic, is the arresting ability to live with strangeness."[35] Moss's reading of Hogg evokes Timothy Morton's coinage of "strange stranger" as a term that addresses "the paradoxes and fissures of identity *within* 'human' and 'animal'" (original emphasis) and insists on the irreducible uncanniness of beings.[36] The overriding effect of "The Surpassing Adventures of Allan Gordon" is of a confusion that emerges in the paradoxes of identity, apparent in the contortion of Hogg's syntax as much as in the excesses of his plot and the intricate transferences between man and bear.

Certainly, no claims could be made for Hogg's story as utopian or idealizing in its depiction of a drunken tailor's febrile negotiations with arctic ecology. But it is precisely the provisionality, uncertainty, and sheer weirdness of "Allan Gordon" that makes it such a salutary corrective to the era's overdetermined nationalist romances of the barren North. Confusion operates as the antithesis to economic calculation, imperial confidence, and masculinist posturing and facilitates what Anat Pick in another context refers to as a "creaturely ethics" that "does not depend on fulfilling any preliminary criteria of subjectivity and personhood."[37] Hogg never himself traveled to the Far North; his text remains an act of imaginative speculation. His strange tale of Allan and Nancy, however, contests a normative imperialist history of arctic exploration and provides a more complex and self-reflexive view on the meanings of encounters between human and nonhuman creatures in the nineteenth-century Arctic.

Endnotes

1. Robert Southey, *Life of Nelson* (London: Collins, n.d. [1813]), 18.

2. Captain A. T. Mahan, *The Life of Nelson: The Embodiment of the Sea Power of Great Britain* (London: Sampson, Low and Marston, 1899), 11.

3. Southey, *Life of Nelson*, 21.

4. James Stannier Clarke and John McArthur, *The Life of Admiral Lord Nelson, K.B.* (London: T. Bensley, 1810), vi.

5. Mahan, *Life of Nelson*, 11.

6. Chauncey Loomis, "The Arctic Sublime," in *Nature and the Victorian Imagination, 1818–1914*, ed. U. C. Knoepflmacher and G. B. Tennyson (Los Angeles: University of California Press, 1977), 96.

7. Loomis, "The Arctic Sublime," 96.

8. Sarah Moss, *Scott's Last Biscuit: The Literature of Polar Exploration* (Oxford: Signal Books, 2005), 95.

9. Jen Hill, *White Horizon: The Arctic in the Nineteenth-Century British Imagination* (Albany: SUNY Press, 2008), 32.

10. Robert G. David, *The Arctic in the British Imagination, 1818–1914* (Manchester: Manchester University Press, 2000), 168.

11. Cited in Francis Spufford, *I May Be Some Time: Ice and the English Imagination* (New York: Picador, 1999), 51.

12. Ibid.

13. Exactly how valuable a northern sea route to the west would have proved is something of a moot point. Spufford, for example, argues that "[s]ince it was already clear that the strait, if it existed, must be at too high a latitude to be easily navigated, hardly any trading advantage could be gained by the nation that contrived to penetrate it" (*I May Be Some Time*, 51).

14. Russell Potter, *Arctic Spectacles: The Frozen North in Visual Culture, 1818–1875* (Montreal: McGill-Queen's University Press, 2007), 38.

15. Spufford, *I May Be Some Time*, 49.

16. Thomas Richards, *The Commodity Culture of Victorian England: Advertising and Spectacle, 1851–1914* (Stanford: Stanford University Press, 1990), 8.

17. Potter, *Arctic Spectacles*, 72

18. Ibid., 12.

19. Ibid., 11.

20. Spufford, *I May Be Some Time*, 51.

21. Penny Fielding, "'No Pole nor Pillar': Imagining the Arctic with James Hogg," *Studies in Hogg and His World* 9 (1998): 51.

22. Spufford, *I May Be Some Time*, 102.

23. Cian Duffy, *The Landscapes of the Sublime, 1700–1830: Classic Ground* (Basingstoke: Palgrave, 2013), 104.

24. James Hogg, "The Surpassing Adventures of Allan Gordon," *Altrive Chapbooks* 2, no. 1 (1987), 33. All subsequent references to this text appear in parenthesis.

25. Sarah Moss, "Romanticism on Ice: Coleridge, Hogg and the Eighteenth-Century Missions to Greenland," *Romanticism on the Net* 45 (2007): 15, accessed January 15, 2014, doi:10.7202/015816ar.

26. Fielding, "'No Pole nor Pillar,'" 52.

27. Ibid., 53.

28. Ibid., 52.

29. Matthew Calarco, *Zoographies: The Question of the Animal from Heidegger to Derrida* (New York: Columbia University Press, 2008), 92.

30. Ibid., 93

31. Ibid., 94.

32. Giorgio Agamben, *Homo Sacer: Sovereign Power and Bare Life*, trans. Daniel Heller-Roazen (Stanford: Stanford University Press, 1998), 5.

33. Ibid., 1.

34. Moss, "Romanticism on Ice," 17.

35. Ibid., 17.

36. Timothy Morton, *The Ecological Thought* (Cambridge, MA: Harvard University Press, 2010), 41.

37. Anat Pick, *Creaturely Poetics: Animality and Vulnerability in Literature and Film* (New York: Columbia University Press, 2011), 193.

Bibliography

Agamben, Giorgio. *Homo Sacer: Sovereign Power and Bare Life*. Translated by Daniel Heller-Roazen. Stanford: Stanford University Press, 1998.

Calarco, Matthew. *Zoographies: The Question of the Animal from Heidegger to Derrida*. New York: Columbia University Press, 2008.

Clarke, James Stannier, and John McArthur. *The Life of Admiral Lord Nelson, K.B.* London: T. Bensley, 1810.

David, Robert G. *The Arctic in the British Imagination, 1818–1914*. Manchester: Manchester University Press, 2000.

Duffy, Cian. *The Landscapes of the Sublime, 1700–1830: Classic Ground*. Basingstoke: Palgrave, 2013.

Fielding, Penny. "'No Pole nor Pillar': Imagining the Arctic with James Hogg." *Studies in Hogg and His World* 9 (1998): 45–63.

Hill, Jen. *White Horizon: The Arctic in the Nineteenth-Century British Imagination*. Albany: SUNY Press, 2008.

Hogg, James. "The Surpassing Adventures of Allan Gordon." *Altrive Chapbooks* 2, no. 1 (1987): 1–57.

Loomis, Chauncey. "The Arctic Sublime." In *Nature and the Victorian Imagination, 1818–1914*, edited by U. C. Knoepflmacher and G. B. Tennyson, 95–112. Los Angeles: University of California Press, 1977.

Mahan, Captain A. T. *The Life of Nelson: The Embodiment of the Sea Power of Great Britain*. London: Sampson, Low and Marston, 1899.

Morton, Timothy. *The Ecological Thought*. Cambridge, MA: Harvard University Press, 2010.

Moss, Sarah. *Scott's Last Biscuit: The Literature of Polar Exploration*. Oxford: Signal Books, 2005.

———. "Romanticism on Ice: Coleridge, Hogg and the Eighteenth-Century Missions to Greenland." *Romanticism on the Net* 45 (2007): 1–17. Accessed January 15, 2014, doi:10.7202/015816ar.

Pick, Anat. *Creaturely Poetics: Animality and Vulnerability in Literature and Film*. New York: Columbia University Press, 2011.

Potter, Russell. *Arctic Spectacles: The Frozen North in Visual Culture, 1818–1875*. Montreal: McGill-Queen's University Press, 2007.

Southey, Robert. *Life of Nelson*. London: Collins, n.d. [1813].

Spufford, Francis. *I May Be Some Time: Ice and the English Imagination*. New York: Picador, 1999.

PART III

notions of north and nation

transnational norths

7

Northern Relations

Colonial Whaling, Climate Change, and the Inception of a Collective Identity in Northern Alaska and the Northern Atlantic

chie sakakibara / OBERLIN COLLEGE

Introduction

It was the gazes of the three figures that captivated my attention. This portrait of an Iñupiaq family was photographed sometime between 1896 and 1913 by Reverend Samuel Spriggs, the head of the Presbyterian missionary, who is now known for his photography of the lifestyle of villagers and the change brought upon them by mission activities (fig. 7.1). The daughter, age ten or so, stands behind her parents. Her facial expression is not as stern as her parents', and she seems itching to voice her feelings as the photographer told the family not to move. The daughter's phenotypical lack of resemblance to her parents made me wonder about her origins.

In March 2010, I found this photograph in the Historical Photograph Collections at the Alaska State Library. The caption revealed that the portrait is of Olemaun, Fannie Keerik, and Qusalgana of Barrow, Alaska. It also stated that Fannie, the girl, was adopted by Olemaun and Qusalgana, and that her biological father was a Portuguese whaler. I shared this image on Facebook, where I have many friends from all over Alaska. "Is anyone related to Fannie?" The volume of response was overwhelming, and the thread was flooded with excitement from the villages of Point Hope, Barrow, and all over Alaska within a matter of a few hours. Ironically,

Fig. 7.1. Olemaun, Fannie Keerik, and Qusalgana of Barrow, Alaska. Alaska State Library, Reverend Samuel Spriggs. PHOTOGRAPHS. POINT BARROW, ALASKA, 1899–1908, ASL-PCA-320-21

through the transnationalized social media network of Facebook, I came to feel that Portugal was very close to Alaska or, rather, to the hearts of many Iñupiat, through this often-overlooked historical connection.

I am a cultural geographer, and my primary research explores how Iñupiat of arctic Alaska perceive and respond to environmental changes incurred by climate change. Specifically, I explore bowhead whaling and how associated social rituals establish the core of cultural identity and ability to adapt to climate variability. The Iñupiat call themselves the "People of the Whales" as they depend on the bowhead whale for sustenance and cultural meaning.[1] Throughout my fieldwork, I have come to realize how the whale provides the cornerstone of cultural resilience for a hopeful future. Indeed, the whale serves as the foundation of the Iñupiaq sense of place and understanding of their environment. My relationship with the Iñupiaq community eventually went well beyond research, as I became an adopted member of a whaling family and cultivated an everlasting friendship with many community members. It did not take long before I wanted to know more about Alaska-Portugal connections, which I was learning about from Iñupiaq families in Barrow and Point Hope. Rather than deploring European heritage as a lingering legacy of colonialism as I would have assumed, some Iñupiat take it as a positive and unifying force of transnationalization and globalization with which they raise awareness of a collective identity. Such reconfiguration of cultural identity is part of a wider discussion of resilience in this time of radical social and environmental transformation and uncertainty in arctic Alaska.

As Sarah Jaquette Ray and Kevin Maier articulate in their introductory chapter, the North has never been an isolated or desolate place. It has been a hearth to kindle the future of northern peoples in a variety of spontaneous ways. In this chapter, I explore how some Iñupiat of northern Alaska attempt to reclaim their connectivity—or, as Barbara Bodenhorn calls it, "relatedness"[2]—with the Portuguese heritage to enhance their cultural prowess in the time of enduring social and environmental disruptions. Here I must acknowledge the historical and ongoing controversies over essentialized discourses of ethnicity and race, which were often used to advance and justify colonial projects.[3] Concepts such as blood quantum and genealogical forms of indigeneity have consistently facilitated assimilation processes. It is not my intention to advocate the hybridity of race as a solution to engage with and better cope with climate change. Rather, it is my goal to highlight a transformability of cultural identity or acceptance of complex cultural identity and its potential development as a backbone of future indigenous solidarity and self-determination.

The term *acculturation* has never done justice to describing any phases of the Iñupiaq history or worldview. As Richard Glenn of Barrow told me, "Iñupiat lives have gone through great change," and their life and culture are based on adapting to constantly changing cultural, political, and environmental climates.[4] In this context, among Iñupiat, experiences associated with acculturation are often not seen as passive reactions to political and environmental colonialism, and they bear the potential to lead the community to a sustainable future through internal renewal and restructure. Reflecting on my conversations with my Iñupiaq collaborators, various episodes from ethnographic fieldwork, and literature review of historical documents, it is entirely reasonable to interpret the adoptions of Euro-American whalers, especially those from Portugal, as an Iñupiaq effort to seek and stress mutual obligation and long-term sustainability through relatedness.

In many indigenous communities, establishing relatedness within and beyond their community has been a survival scheme,[5] and it has been an important quality to keep their society strong, healthy, and flexible. Studying the Lakota, Raymond DeMallie discusses the importance of transforming strangers into community members "in order to have a means of dealing with them and a basis for trusting them."[6] This strategy allowed them to take better advantage of any potential benefits these newcomers could bring with them.[7] Ella Cara Deloria, Yankton Dakota ethnographer and educator with English, French, and German heritages, called this process "social kinship," which was considered by her Dakota relatives as legitimate as genealogical kinship.[8] As long as the adopted former stranger observed the local custom and traditions, he or she was not different from the rest of the community members.[9] This kinship structure was explicitly revealed among Iñupiat, as Bodenhorn[10] elaborates how historically "being Iñupiaq" has not been about boundary maintenance but rather has been about boundary permeability. Bodenhorn defines an "Iñupiaq" as someone who has a commitment to subsistence activities, especially whaling. The act of sharing, which allows them to create a bond with other communities, is also an important quality to be an Iñupiaq.

Historically, Iñupiat have been incorporating a steady flow of people into local families and social relations. Iñupiat adopted Portuguese strangers and converted them into kin, sharing everything in order to increase their own prosperity and power through alliance, and innovating relentlessly but with respect for time-tested cultural values and models of appropriate conduct. It is possible to infer that in the late nineteenth and early twentieth centuries, Iñupiat saw Portuguese whalers as relatives, not a force of colonialism or a superior being from another world, and that they had their own positive agenda in agreeing to incorporate them into their social fabric for their mutual benefit. Another dimension of this development is that, as many of my Iñupiaq acquaintances noted, the majority of the "Portuguese" whalers in Alaska were actually from islands in the northern Atlantic such as the Azores and Cape Verde.[11] While labeled "Portuguese," the Azorean and Cape Verdean islanders bear significant cultural distinctions and historical experiences, which were amplified by the effort to intentionally distance themselves from those of the continental Portuguese. Welcoming the men who kept the distinctive Azorean or Cape Verdean spirits might have been seen as a chance to develop a useful alliance with possibly powerful outsiders and thereby enhance the outside world's respect for Iñupiat. What a colonial wave brought to northern Alaska, in other words, was actually desirable for those Iñupiat who hoped Portuguese would help them to fortify their sense of place and community with an enhanced degree of reciprocity.

This chapter, therefore, challenges our perception of the "North" by confirming the historical and contemporary linkage of the western Arctic and northern Atlantic, connecting environments, histories, and peoples of the North with the spaces and communities that are (traditionally perceived as) non-northern. Illustrating the contemporary manifestation of cultural identity between the two distinct cultures and peoples, this chapter will reveal how their heritages and identities may blend in the northernmost portion of Alaska today, and how this process may transmit to their relatives in the northern Atlantic as a form of shared collective identity for the future.

While this study draws on historical sources and contexts, I strive to make it a dynamic account that facilitates the understanding of the past, contextualizing a missing link, and assisting today's Portuguese descendants in Iñupiaq Alaska and their Portuguese relatives in reconfirming the heritage and culture they live today. This study shows how place-based identities are not necessarily a defensive reaction to globalization but might be a product that reveals a new solidarity generated by the desire for developing cultural resilience.

It is my hope that this study contributes to the literature of decolonization through the lens of indigenous geography. Reconceptualization of the world and ethnic and cultural heritage often serves as an example of decolonizing efforts. Maps have been used as the main tool to colonize the world, but they can also serve as a means for decolonization through a strong effort of physical and imaginative countermapping, as was exemplified by the indigenous activists and their use of sketch maps to reclaim territories in Kalimantan, Indonesia,[12] as just one of many examples. Turning to arctic Alaska to study the significance of development of collective identity, we might learn lessons and insights for communities elsewhere that are going through similar issues incurred by socioenvironmental changes and challenges.

Portuguese Whalers in Iñupiaq Alaska

The lineage of Euro-American commercial whalers entered the Iñupiaq community in the second half of the nineteenth century and gradually increased its influence upon the society.[13] By the time commercial arctic whaling became a well-established enterprise, a large stretch of arctic Alaska had become a crossroad of exchange and culture, where members of different Alaska Native groups and Euro-American whalers and traders interacted regularly.[14] As many of today's Iñupiaq family names indicate, many whalers who married Iñupiaq women were of English origin via New Bedford, Massachusetts, one of the most prominent whaling hubs in the world in the nineteenth century. However, the Portuguese whalers also played an instrumental role in American commercial whaling in the Arctic and also brought whalers of African descent to Alaska. Furthermore, a connection between the Portuguese and Alaska Natives can be verified by the existence of the pidgin language that was used for communication between the commercial whalers and the Iñupiaq whalers, which was "a mixture of Eskimo, English, Portuguese, and South Sea Island words, and has been evolved by the whalers in their many years of association with the natives."[15]

Within each ethnic group, individuals' characters, experiences, and attitudes affected relations with outsiders, as did the circumstances in which they met.[16] Portuguese whalers and Iñupiat met within larger contexts of cultural collision and colonial encounter that originally governed their interactions. Despite Shoemaker's thesis that Native American whalemen did not behave differently from their Euro-American counterparts during the whaling expeditions,[17] the fact that the majority of the Portuguese whalers in Alaska actually originated from islands claimed by Portugal has likely made a significant impact on their perception of themselves in relation to indigenous communities outside of New England. In other words, the "Portuguese" whalers might have held a certain degree of affinity and sympathetic feelings toward the local community in arctic Alaska. Portuguese explorers discovered and claimed the uninhabited Cape Verde Islands in the fifteenth century, and the islands became the focal point for the Atlantic slave trade. The Azores were uninhabited until the age of discovery, and were settled shortly thereafter by immigrants from Portugal and Flanders who had fled Europe because of plagues, war, starvation, crime, rebellion, and overpopulation. However, overpopulation on the islands soon led to starvation and lack of employment. In the mid-nineteenth century, commercial whalers from New England stopped often in the Azores to recruit replacement crew members. Around the time when many commercial whaling vessels came from New England, Pico—which later became the major supplier of young whaling crewmen for the United States—was an island of great poverty. The continuous volcanic eruptions and a persistent grapevine disease caused famine, all exacerbated by feudal ownership of land throughout many of the islands. The men were forced to turn to the sea for their livelihood. Being successful whalers helped the islanders construct their own cultural identity in opposition to Portuguese dominance. In fact, in the nineteenth and twentieth centuries, many Azorean youths had taken advantage of being whalers to illegally migrate to America to acquire wealth and to eventually return to their homeland to support or to begin a family.[18]

Whaling voyages would last three to five years with minimum wages or no compensation other than transportation to the United States or Canada. At the end of the voyages, many settled in New England, California, and Hawai'i, creating an Azorean diaspora. In Alaska, a number of Azoreans and Cape Verdeans eventually settled at shore-based stations near Point Hope and Barrow, two prominent Iñupiaq villages. Ironically, they were collectively called "Pulgi" (Portuguese) in the Iñupiaq language, and several whalers such as Joe A. Ferreira and Antonio Bettencourt "went native," became adopted people, and married or lived with Iñupiaq women in long-term partnerships, simultaneously developing kin relations with village men.[19] This is the origin of the multilayered cultural and ethnic identity embraced by some Iñupiat today.

Transforming Historical Trauma into a Collective Identity

Like other indigenous peoples in the world, Iñupiat have experienced profound historical trauma and disruptions, including epidemics, displacement, forced establishment of permanent settlements and resettlement, cultural and economic colonization, and religious conversion since the advent of Euro-American explorers, commercial whalers, and Christian missionaries starting in the mid-nineteenth century. The advent of the Western/southern outsiders were followed by the implementation of the boarding school system, which sent Iñupiaq children to distant boarding schools to facilitate the assimilation process. These historical events have caused acculturation stress and identity conflicts, and rapid social change created significant health and social problems among them.[20] Recent environmental degradation and the consequent disruption of traditional lifeways have exacerbated critical social problems facing indigenous populations in the circumpolar Arctic:[21] alcoholism, drug abuse and addiction, suicide, depression, domestic violence, and low motivation.

However, many Iñupiat—both young and old—also recognize the benefits of connecting and relating to the world, and they do so sometimes through their Euro-American heritage. Instead of seeing this connection as an extension of colonialism and capitalism, Iñupiat see globalization as a force to foster a new form of collective identity, in addition to their place-based understanding of self. Working closely with his Blackfeet collaborators, Malcolm McFee[22] shows that acculturation need not mean cultural loss or replacement. Rather, new ways and values can be learned without sacrificing the original worldview, and this incorporation makes a culture more elastic and resilient, making someone "150% man." The Iñupiat with Euro-American heritage see themselves as no less Iñupiat than others.[23] The Portuguese ancestry is seen as a valuable resource to tie Iñupiat with their past, present, and future. Some community members of Point Hope and Barrow are now more willing to accept and advance identities that do not fit neatly into mutually exclusive categories, challenging essentialized indigenous stereotypes.

Resilience and Collaborative Reciprocity

Many studies have confirmed how strong correlations between positive affiliation and engagement with traditional culture form social well-being and resilience.[24] Resilience consists of the processes by which people overcome life challenges to achieve their sense of well-being and is often deeply rooted in their sense of place.[25] The connection between culture and these processes is clear, and there has been some scholarship on history and memory that emphasizes connections between historical understandings and cultural identity. However, few studies have explicitly revealed how cultural identity and its development play into indigenous social well-being and resilience.[26] Therefore, it is necessary to explain how a strong and positive synthesis of its culture, history, and even influences from the outside world sustains a community when it responds to social and environmental hardships.

Resilience is the capacity of an individual, entity, or organization to maintain and renew itself when its existence or viability is challenged or threatened. The term *resilience* can be used with many different modifiers, such as ecological resilience, cultural resilience, and psychological resilience. The Iñupiaq culture is an example of cultural resilience because they have successfully survived in the northern rim of the world by retaining and building upon their subsistence lifestyles for over a thousand years. In other words, Iñupiat and their northern circumpolar relatives are prime examples of human adaptation to changing physical and social environments.[27] Anthropologist Joseph Jorgensen observes and describes the Iñupiaq community: "Eskimo villages are resilient places" and Iñupiat had no difficulty in synthesizing newly introduced commercialism and dependencies with traditional subsistence-based lifestyles.[28] While the political and economic currents since 1970 have modernized the villages and integrated Iñupiat into capitalist economy by enhancing their reliance on the public and private sectors of the national economy, "there is a determination on the part of Eskimos to maintain traditional Eskimo culture and at the same time to adopt a pragmatic acceptance of the benefits of modern technology."[29] Bodenhorn[30] also points out the importance of balancing the old and new elements of life among the Iñupiat.

Similarly, the notion of "collaborative reciprocity"[31] is crucial for understanding the northern indigenous resilience as reflected in traditional whaling. Central to the Iñupiaq notion of collaborative reciprocity is the belief that humans and animals physically and spiritually constitute each other, that the soul, thoughts, and behaviors of animals and people interpenetrate in a collaboration of life. As my field experience also testifies, this relationship is fundamental to many indigenous cultural and environmental ethics, a cornerstone for an appropriate way of life.[32] For example, many indigenous groups in the Arctic understand that animals willingly give themselves to hunters in response to receiving respectful treatment as "non-human persons."[33] Such relations are understandably influenced by animal behavior, distributions, and availability, all of which are conditioned by the environment, which is in turn influenced by climate and climate change.

Collaborative reciprocity also should be understood as an adaptive strategy that engages arctic life. Adaptation is a key to resilience for the community in that it enables the renewing and reconsidering of their way of life. Not only does it link human welfare with that of animals, it affords people a means of expressing their concerns about environmental changes and for coping with, adapting to, and living through such changes. This interrelationship is demonstrated in various ways, most of them publicly visible. In Iñupiaq society, collaborative reciprocity is revealed in the way whale hunting is organized and conducted, the manner in which food is prepared and consumed, in the use of whale body parts in the built environment, in the content and performative styles of Iñupiaq music and storytelling, and in many other aspects of ceremonial and everyday life.

In order to explore Iñupiaq resilience further, I would like to expand this understanding of collaborative reciprocity. The indigenous populations in the Arctic face critical social problems that are also linked with the colonial legacy and recent environmental degradation that threatens the future of the community. Historical trauma consists of cumulative stress caused by assimilation and acculturation, loss of culture, and experience and history of genocide and racism.[34] Such continuing and cumulative trauma has its roots in colonialism and influences the current and future well-being of indigenous communities. The significantly higher youth suicide rate for indigenous populations compared to other minority groups especially exemplifies the plight of the communities.[35]

At this time of heightened socioenvironmental stress and internalized and institutionalized oppression, Iñupiat increasingly rely on the power of cultural solidarity through traditions.[36] Solidarity as a form of resilience can also be found in hunting, religious worship, language retention efforts, and a variety of manifestations of expressive culture such as arts, music, and dance along with heritage education. In a contemporary context, music has a power to resuscitate cultural and collective memories among Iñupiat, and these memories have an important role in the community's responses to environmental uncertainties and their related emotional difficulties.[37] In northern Alaska, dancing, singing, and drumming are powerful cornerstones for cultural survival. In particular, dance reveals a strong sense of social integration and moral development. As Aaron Fox writes, "[d]ancing saves lives"[38] as much as sharing stories does.[39] Expressive culture exemplifies a form of collaboration with humans and nonhuman beings as well as the reciprocity between different age groups, genders, and other communities within and beyond the circumpolar region. In other words, shared heritage plays a vital role in having people redefine who they are by facilitating exchange and communication between people, as well as between people and environment, integrating the community into the structures of material and emotional reciprocity. In this way, heritage appreciation and reconfirmation may serve as an emblem of cultural empowerment and sovereignty. My experience and observation in Alaska often confirmed how accomplishing these tasks fosters crucial affirmation of cultural identity.

Whaling as a Cultural Emblem

One's culture and identity often become most visible when one confronts "others." When encountering difficult values and beliefs, the taken-for-granted features and customs of culture help people justify themselves and better communicate with outsiders.[40] Cultural exchange and cross-cultural interactions oftentimes incur a sense of cultural and identity crisis, and when our worlds are felt to be endangered, "traits" may turn into "emblems" or cultural emblems, emotionally charged markers that we use to construct boundaries and barriers in self-defense.[41] Once transformed into emblems, the traits serve as an irreplaceable foundation to firm up our unity as one people to communicate with dominant others. Both Jean Briggs and Louis-Jacques Dorais elaborate that defining an ethnic and cultural identity in the Arctic is a recent phenomenon among the Inuit, which is also true among the Iñupiat.[42] Briggs discusses possible emotional origins for emblems that Canadian Inuit often use in differentiating themselves from Euro-Canadians. Emblems can take the form of materials, behaviors, and/or institutions that may support one's culture and cultural well-being. In discussing cultural and ethnic identities of Inuit in Canada, Nobuhiro Kishigami[43] has shown how indigenous institutions anchored in past practices sustain the continuity of Inuit cultural traditions and foster a sense of being Inuit in multiethnic settings of contemporary Canada.

While their goal seems to be contradictory to their predecessors' efforts to differentiate themselves from Euro-Canadians/Americans, the Iñupiat who claim Portuguese ancestry are turning the notion of colonial legacy—including history and trauma—into a positive emblem of self-determination to enhance their connectivity among themselves through their relatedness with the "northern Atlantic" People of the Whales. In their case, globalization enhanced cultural identity by highlighting the sense of self as well as the connection between the community and the world.[44] Despite the differences of the whale species between the two regions, it is intriguing that both Iñupiat and Azoreans recognize the whale and whaling as their cultural emblems. Below I will highlight how Iñupiat and Azoreans respectively attempt to cope with socioenvironmentally induced stresses by strengthening their self-representations as the Peoples of the Whales.

Cetaceousness in Arctic Alaska

"Jabbertown was a home to different people, so we said. They spoke different languages, [had] darker skins, [were] taller and stronger. I hear they were very powerful people. Some Eskimo women married these men who came with whaling vessels. The men were [mostly] Portuguese." Describing Jabbertown, Suuyuk Lane, an Iñupiaq elder then in her early eighties, shared her memories: Jabbertown is located approximately ten miles southwest of today's Point Hope, Alaska, the northwesternmost settlement of North America. Point Hope, also known as Tikiġaq in Iñupiaq, holds approximately eight hundred people, most of whom are indigenous Iñupiaq. Jabbertown started out as the Iñupiaq hunting ground of the bowhead whale from a traditional whaling seal-skin boat called *umiaq*. Portuguese whalers initially arrived in the mid-nineteenth century on American vessels with commercial whalers from New England and California. The residents of Jabbertown engaged in trading for skins and other local goods with the community at Point Hope. As Mark S. Cassell notes,[45] commercial whaling did not destroy the fabric of Iñupiaq life and the cultural cores despite many unexpected and disastrous issues that the indigenous people had to confront. Other Portuguese continued to move in and out of Jabbertown during these years, resulting in interracial marriages between Portuguese whalers and Iñupiaq women.[46] Today, many families in Point Hope recognize their Portuguese ancestry and, indeed, more than a few bear family names that originated in Portugal although they are today mostly Anglicized. "The curly hair is a proof that the whales put Pulgi and Eskimos together," said Suuyuk. "Now they are all our relatives."

This integral relationship with the bowhead whale that ultimately united Iñupiat and Pulgi heritages is now being jeopardized. Iñupiat are especially susceptible to climate and associated environmental changes because they rely on sea ice to hunt the bowhead whale. Despite social change over the last century the bowhead remains central to Iñupiaq life.[47] Through ethnographic fieldwork, I learned how a whalecentric worldview based on the notion of collaborative reciprocity—what I call *cetaceousness*—has been influenced by unpredictable environmental change and how people work toward retaining their identity through their physical and spiritual associations with the whales. Cetaceousness is a hybrid of *cetaceous* and *consciousness*, which links human awareness with cetaceans or whales. The idea of cetaceousness evokes the strong interrelatedness of Iñupiat and the whales at both spiritual and physical levels.

Cetaceousness is a social and emotional process that Iñupiat use to communicate with the whales and vice versa. It is a means of grasping an entire epistemology and native sense of place. Among the Iñupiat of Point Hope, an origin story portrays the dramatic transformation of a bowhead into their ancestral homeland.[48] To Iñupiat, then, their land used to be a whale whose death coincided with the birth of a human habitat. Furthermore, human and whale identities are interchangeable in Iñupiaq storytelling, which exhibits emotional kinship among the people, the land, and the whales.[49] Negative changes in human-whale relations therefore will cause a substantial shift in Iñupiaq well-being and cultural continuity.

While my Iñupiaq correspondents unanimously expressed the importance of unity between humans and whales, and as an integral part of their epistemology, I soon learned that there were no Iñupiaq concepts or words directly equivalent to cetaceousness, except possibly the word *Iñupiaq* itself, since it also means the People of the Whales. During my conversations with villagers, I casually brought up the idea of cetaceousness—I knew the term does not quite work linguistically in English—to see how they would respond and if they had other words or ideas associated with it in their cultural tradition or vocabulary. Isaac Killugvik of Point Hope told me that he did not recognize the importance of coming up with such a term when he was younger to communicate with outsiders to explain their relationship with the whales. However, "now that we have strange weather, late winters and early springs, and who knows something bad may happen to the whales," continues Isaac, "we may need that term [cetaceousness] to have non-Eskimo people understand the value of the whales to our lives and urge them to think about our sea and land." Ben Nageak of Barrow describes his people's fundamental relationship with the bowhead as "the *whale*-ness of the Iñupiat." The whale is deeply intertwined with human existence and spirituality. As Eugene Brower commented, "*Aġviiḷuta Iñuuyumiñaitchuġut* [if there are not any whales, we are not here]." Studying the transformation and development of Iñupiaq cetaceousness in the face of climate change allows us to learn from Iñupiaq culture and the vitality of its resilience.

In the face of contemporary global challenges, the recent Iñupiaq movement to revive the Portuguese heritage reveals an additional dimension of cetaceousness to consolidate and reconfirm the kinship within and beyond their own community. The following responses to the family portrait of Fannie Keerik that I shared on Facebook encapsulate the sense of excitement and hope that the descendants of Portuguese Iñupiat embrace for the future:

> "Yes, she is our amau [great-grandmother]. What a wonderful photo. . . . It would be interesting to hear about the Portuguese whalers as many of us are related to them!"

> "I have always mentioned to my family that I want to go to Portugal once I heard that my great-great-grandfather has originated from there. I think we should go as a family once we know more, Portuguese is part of us now."

> "That is our amau, I tracked our great-great-grandfather from the Azores and Cape Verde. But I kept going in circles and couldn't go any further. A few of us have been doing research on Fannie's father but it is hard."

> "Fannie is also my amau and I often think of our heritage from Portugal from her side of the family would like to hear more sometimes."

> "Fannie is my grandmother's grandmother. The adoption part is interesting. . . . For me to know that I am a descendant of a Portuguese whaler."

> "This heritage is very important to us. So, where do we go from here? What shall we do next? We should keep talking, and we'd love to meet the Portuguese people."

Açorianidade in the Northern Atlantic

The Azores enjoy a mild climate throughout the year except for occasional winter storms when deep Atlantic depressions track across the islands. Since the initial immigration, settlers have arduously cultivated the dark and rocky volcanic soil. As a result, the Azores have served as a bread basket of Portugal, particularly for wheat, wine, and dairy products. Currently the status of the islands is an Autonomous Region of Portugal, but its geographical distance and unique history guarantee an identity distinct from Portugal.

With the photograph of Fannie Keerik and the stories that my Iñupiaq collaborators and friends shared with me in northern Alaska, I flew to Pico of the Azores for the first time in 2010.[50] Unlike the common knowledge that the people held in Alaska regarding the strong historical bond between Iñupiat and Pulgi, the fates of the Portuguese whalers who ended up in Alaska were not widely known. "Do we have relatives in Alaska?" On the visit, I met many individuals who were intrigued with the fate of their long-lost relatives from the past. Not surprisingly, the social and environmental problems that are central to the islanders are very similar to the issues that their Alaska relatives face. Today, as much as in the Arctic, as the effects of climate change take hold among Azorean communities, the role and place of whales in sustaining identities become greater as a vital cultural emblem. Cultural response to climate change among the Azoreans is simultaneously both similar to and different from their Alaska counterparts. For the Azorean islanders, in spite of the relatively recent adoption of whaling in the nineteenth century, the sperm whales, *cachalote* in Portuguese, not only provided economic sustenance but also nourished human emotions and expressive culture and helped the people cultivate their sense of place.[51] Undoubtedly whaling and its legacy have contributed to the growth of the modern Azorean social, cultural, and religious life. As the recent threat and fear brought to the islanders by Hurricane Alex exemplify, the islanders are now experiencing higher unpredictability of weather and climate, and a future scenario predicts sea-level rise, frequent flooding, and erosion that would change the fate of all the islands' relation with the ocean.

The notion of *Açorianidade* (Azoreanness) refers to conscious and unconscious efforts of the islanders to construct a distinctive cultural awareness of their own.[52] *Açorianidade* was never static but transformational because of the islands' geographical locations and accessibility to the rest of the world. The historical landscape paintings of the Islands of Faial and Pico by American painter Benjamin Russell (1841), now housed in the Old Dartmouth Society–New Bedford Whaling Museum, reveal the fact that the Azoreans were always exposed to foreign influences, a fact

Herman Melville illustrated in *Moby-Dick* (1851). Many port towns in the Azores grew as crucial commercial ports for the Atlantic sailing route connecting Europe, India, China, Japan, and Brazil. Just like the Cape Verde and Canary Islands, the Azores connected the Old and New Worlds via oceanic trading networks, which ultimately reached the High Arctic.

As much as the deep-sea whaling facilitated by the American whaling companies became popular among the islanders, coastal whaling also started growing its roots in the Azores. The year 1832 marks the beginning of shore whaling in the Azores.[53] Feasts and celebrations often followed a successful hunt. In the Azores, a new genre in literature appeared with the introduction of whaling: short novels based on whaling were written by Nunes da Rosa (1871–1946), Rodrigo Guerra (1862–1924), Florêncio Terra (1859–1941), and Dias de Melo (1925–), just to name but a few. The last whaling ship from the United States of America left the Azores in 1922, which also ended an almost centurylong Portuguese migration to the United States. However, by then off-shore whaling based on the Azores had become an integral part of the islanders' economy and subsistence calendar, and the symbolic importance continued well into the 1980s. The whaling activity itself was eventually transformed into their cultural emblem beyond its termination.

The whaling tradition in the Azores came to a sudden halt in 1987. For the protection of the cetacean species, the International Whaling Commission (IWC) banned whaling and the Portuguese government agreed without providing sufficient compensation and explanation to Azorean communities. The whalers, especially elderly ones, firmly resisted the decision. As a form of protest in November 1987, Pico whalers hunted three sperm whales off Lajes. They invited television crews and journalists to have them broadcast the hunt as they launched out to sea. Shortly after, Portugal officially joined other nations in an agreement that prohibits the killing of the whales. This is how whaling, the emblem that once distinguished *Açorianidade* from Portuguese, ceased to exist in this way. Since 1988, major whale factories in the Azores have become whaling history museums.[54]

Despite this, the whale remains the emblem of cultural survival for the Azoreans. Manuel da Costa, director of the Whalers' Museum, who is a descendant and namesake of a renowned whaling captain, says: "The whales still represent *Açoreanidade* and a trademark of the Azores. We should all agree that the image of the Azores to the international arena is the 'tail of the whale' [a metaphor of the "Hand of God"]. In the past, people hunted the whale, which formed the basis of their relationship, but now the whale forms a new relationship with the people." The transformation from the "Whaling People" into the "Whale-Watching People" certainly brought a new power relationship between the people and the whales.

Anthropogenic climate change has begun taking a toll on the Azorean environment. Island life is based on change, and the islanders are currently experiencing a lot of it. Placed in the midst of the northern Atlantic, the Azores is an archipelago that is vulnerable and sensitive to many changes in the sea, the land, and the atmosphere. The main reason for this sensitivity has to do with the limited landmass, where relocations of residents, animals, and resources cannot happen easily.[55] Worse, islands are generally highly dependent on their coastal areas, a major target of climate change impacts.[56] This is the place where human-whale interactions take place.

In June 2009, in hopes of vocalizing itself at the United Nations Climate Change Conference in Copenhagen (COP 15), the Azorean Regional Government founded the Commission for Climate Change in the Autonomous Region of the Azores (ComCLIMA), under the jurisdiction of the Regional Secretariat for the Environment and the Sea, to work on a proposal of the plan for climate change. The ComCLIMA also aims to provide technical and scientific monitoring, information, and consultancy to the Regional Secretariat for the Environment and the Sea regarding the various national and international policies on climate change. More recently, in looking to raise their visibility in the international context and in advance of COP 21 in Paris, the Regional Secretary for the Environment of the Azores declared that nearly a quarter of its electricity is produced from renewable sources, and that the Regional Government expects to increase the production to three quarters by 2018. In order to accomplish this goal, the Regional Government counts on the support of the regional, national, and international scientific systems. They have signed a protocol with the Foundation for Science and Technology, MIT Portugal, the University of the Azores, and the Green Islands project in order to develop a sustainable energy system. With all these measures, the Regional Government is declaring its energy autonomy. This movement shows the potential competitiveness in sustainable development as a factor to consolidate *Açorianidade* in relation to mainland Portugal. In Portugal, dramatic transformations of politics and culture over the last three decades have introduced new dynamics into how traditional culture has been used to assert regional and national identity,[57] and the recent growth of *Açorianidade* may exemplify such a trend. Once, the sperm whale was the major resource provider in the Azores. As the keepers of the whale, the islanders have been working to accomplish their cultural resilience. The recent energy autonomy movement in the Azores exemplifies how the originally whale-based cultural autonomy is taking a new turn in response to globalization, politics, and changing environment. While the whales no longer supply the natural resources for the community, they still remain at the center of the people's cultural and environmental worldview.

Conclusion: A Shared Future of Hope

As I was putting the finishing touches on this chapter, Hurricane Alex was churning northward across the Atlantic Ocean toward the Azores. Hurricanes in the Atlantic in January are rare (the last time it happened was 1938), and the islanders were concerned because a long and wet winter had saturated the soil, potentially leading to large-scale landslides. Meanwhile, I kept hearing the news that sea ice conditions in the Arctic Ocean had not improved since the fall, which made the residents of Barrow and Point Hope anxious as the whaling season was fast approaching. As of September 11, 2015, the arctic sea ice extent was the fourth lowest in the satellite record, reinforcing the long-term downward trend. Climate change now intimately influences the well-being and identities of people in both northern Alaska and the Azores, and this sense of crisis reinforces their connection with the whales to sustain their cultural prowess. In their relationships to the land and to one another, the Azorean islanders often expressed more in common with Iñupiat than with the continental Portuguese. Both Iñupiat and Azoreans are known for their attachment to their homeland, and they express their rootedness to land in similar ways, such as "I grow out of this land" and "the whale is the foundation of our land."

In 2008, an Iñupiaq delegation of educators, hunters, and performers attempted to reestablish their linkage with their Portuguese ancestors through the New Bedford Whaling Museum. They also envisioned conserving the historical collection of Iñupiaq artifacts, including an *umiaq*. Coincidentally, about the same time a group from the Azores organized by the Whalers' Museum in Lajes do Pico had also visited New Bedford to explore the history of Portuguese-American commercial whaling. Unfortunately, the two trips did not overlap. The two groups, however, both clearly see the importance of reestablishing their relationship to commemorate the past, celebrate the present, and explore the future collaboratively and transnationally. These efforts include cultural exhibitions of the People of the Whales aimed at visitors to the Iñupiat Heritage Center in Barrow, as well as participation in an Education through Cultural and Historical Organization (ECHO) grant sponsored by the National Endowment for the Humanities, 2005–2013. This ECHO project united the people in northern Alaska, Hawai'i, and New Bedford to promote cultural exchange among peoples with a shared whaling heritage.[58] The three groups of people (the Portuguese whalers also played a vital role in whaling in Hawai'i; as for the Hawaiian loanwords in the Arctic, see Drechsel and Makuakāne[59]) have retained their ties with the whales in a variety of ways. Throughout the duration of the grant, they organized annual events to facilitate cross-cultural understanding on whale-based subsistence and worldviews, and these activities were only the beginning of their partnership. These reciprocal and intellectual exchanges were tied to broader expressions of cultural and political autonomy and decolonization efforts.

While it is still at the very early stage of its development, having this vision is a vital step for the future of the two Peoples of the Whales. Undoubtedly whaling and its legacy have contributed to the growth of modern Iñupiaq-Azorean social and cultural life. Azoreans are now experiencing higher unpredictability of weather and climate but refuse to be the victims of environmental colonialism. They are actively in search of a way to adapt by retaining their emotional kinship with the whales, people, and places and are eager to learn from their arctic relatives, who are on the frontline of global climate change. It is the whale that brought the two cultures together once again, this time in an environmental context of decolonization and imaginary countermapping through the relatedness. In Alaska and the Azores, the whale provides the critical cornerstone of collective cultural resilience, which people embrace to imagine a hopeful future by bringing the two groups of the People of Whales together once again.

Acknowledgments

Many people and institutions graciously assisted me in making this research possible and productive in both Alaska and the Azores. My fieldwork in the Azores and Lisbon, Portugal, was funded by the McColl Family Fellowship offered through the American Geographical Society. My research was supported by Columbia University, National Science Foundation, Iñupiat History, Language, and Culture Commission, and the Whalers' Museum in Lajes do Pico in the Azores. I owe a great debt to John R. Bockstoce, former director of the New Bedford Whaling Museum and scholar of whaling and arctic studies. The New Bedford Whaling Museum's exhibition on Azorean historical whaling activity was particularly insightful during the early phase of my research on this topic. Onésimo Teotónio Almeida of Brown University and Adelina Azevedo-Axelrod of the John Carter Brown Library and the New England Azores House generously assisted me in connecting with people in the Azores. Donald Warrin of the Bancroft Library Regional Oral History Office at the University of California, Berkeley, kindly shared his time in answering my questions about Portuguese whalers in Alaska. On Pico, Manuel Francisco da Costa, director of the Whalers' Museum, welcomed me to the Whalers' Museum in Lajes do Pico. Ana Fagundes Neves generously and patiently spent days with me to share the stories and history of her people on Pico. Kyle Powys Whyte of Michigan State University has generously shared his expertise in indigeneity and U.S. settler colonialism and associated discourses with me. I am also grateful to the support and inspiration provided by Janet Fiskio and other colleagues in the Environmental Studies Program at Oberlin College. Last but not least, continuous rapport from my friends and collaborators in northern Alaska has generously poured the whale oil into my heart—special thanks are due to Suuyuk Lane, Julia and Jeslie Kaleak, Sr., and the people of Point Hope and Barrow, Alaska, many of whom are proud of their Portuguese ancestry. *Obrigadíssima* and *Quyanaqpak.*

Endnotes

1. Chie Sakakibara, "No Whale, No Music: Contemporary Iñupiaq Drumming and Global Warming," *Polar Record* 45, no. 4 (2009): 289–303; Sakakibara,"Kiavallakkikput Aġviq (Into the Whaling Cycle): Cetaceousness and Global Warming among the Iñupiat of Arctic Alaska," *Annals of the Association of the American Geographers* 100, no. 4 (2010): 1003–1012; Sakakibara, "Climate Change and Cultural Survival in the Arctic: Muktuk Politics and the People of the Whales," *Weather, Climate and Society* 3, no. 2 (2011): 76–89.

2. Barbara Bodenhorn, "'He Used to Be My Relative': Exploring the Bases of Relatedness Among Iñupiat of Northern Alaska," in *Cultures of Relatedness: New Approach to the Study of Kinship*, ed. J. Carsten (Cambridge: University of Cambridge Press, 2000).

3. A. Quijano, "Colonality and Power, Eurocentrism, and Latin America," *Nepentla: Views from the South* 1, no. 3 (2000): 533–580; P. Spruhan, "Legal History of Blood Quantum in Federal Indian Law to 1936," *South Dakota Review* 51 (2006): 1–50; K. TallBear, "Narratives of Race and Indigeneity in the Genographic Project," *Journal of Medicine Ethics and Law* 35 (2007): 412; K. P. Whyte, "Indigeneity and US Settler Colonialism," in *Oxford Handbook of Philosophy and Race*, ed. Naomi Zack (Oxford: Oxford University Press, 2016).

4. Bodenhorn, "'He Used to Be My Relative'"; C. Wohlforth, *The Whale and the Supercomputer: On the Northern Front of Climate Change* (New York: North Point Press, 2005).

5. Bodenhorn, "He Used to Be My Relative"; Bodenhorn, "Is Being 'Iñupiaq' a Form of Cultural Property?," in *Properties of Culture—Culture as Property, Pathways to Reform in Post-Soviet Siberia*, ed. E. Kasten, 35–50 (Berlin: Dietrich Reimer Verlag, 2004); Sergei Kan, "Friendship, Family, and Fieldwork: One Anthropologist's Adoption by Two Tlingit Families," in *Strangers to Relatives: The Adoption and Naming of Anthropologists in Native North America*, ed. S. Kan (Lincoln: University of Nebraska Press, 2001).

6. R. J. Demallie, "Kinship and Biology in Sioux Culture," in *North American Indian Anthropology: Essays on Society and Culture*, ed. Raymond J. DeMallie and Alfonso Ortiz (Norman: University of Oklahoma Press, 1994), 131.

7. M. E. Harkin, "Ethnographic Deep Play: Boas, McIlwraith, and Fictive Adoption on the Northwest Coast," in *Strangers to Relatives: The Adoption and Naming of Anthropologists in Native North America*, ed. Sergei Kan (Lincoln: University of Nebraska Press, 2001).

8. E. Deloria, *Speaking of Indians* (New York: Friendship Press, 1944).

9. Kan, "Friendship, Family, and Fieldwork."

10. Bodenhorn, "'He Used to Be My Relative.'"

11. See more information in Tom Lowenstein, *Ultimate Americans: Point Hope, Alaska: 1826–1909* (Fairbanks: University of Alaska Press, 2008), and Don Warrin, *So Ends This Day: The Portuguese in American Whaling 1765–1927* (North Dartmouth: University of Massachusetts Dartmouth Center for Portuguese Studies and Culture, 2010).

12. N. L. Peluso, "Whose Woods Are These?: Counter-Mapping Forest Territories in Kalimantan, Indonesia," *Antipode* 27, no. 4 (1995): 383–406.

13. John Bockstoce, "Eskimo Whaling in Alaska," *Alaska Magazine* 43, no. 9 (1997); Bockstoce, *Steam Whaling in the Western Arctic* (New Bedford, MA: Old Dartmouth Historical Society, 1977); Bockstoce, "History of Commercial Whaling in Arctic Alaska," *Alaska Geographic* 5, no. 4 (1978); Ernest S. Burch, *The Traditional Eskimo Hunters of Point Hope, Alaska: 1800–1875* (Barrow: North Slope Borough, 1981.)

14. Nancy Shoemaker, "Whale Meat in American History," *Environmental History* 10, no. 2 (2005): 269–294; Shoemaker, *Native American Whalemen and the World: Indigenous Encounters and the Contingency of Race* (Chapel Hill: University of North Carolina Press, 2014); Warrin, *So Ends This Day.*

15. Harold Noice, *With Stefansson in the Arctic* (London: George G. Harrap and Co., Ltd., 1924), 71.

16. Shoemaker, *Native American Whalemen and the World.*

17. Ibid.

18. Warrin, *So Ends This Day.*

19. Charles Brower, *The Northernmost American: An Autobiography*, typescript, Dartmouth College Library Stefansson Collection (n.d.); Don Warrin, interview transcript by Don Warrin with Elsie Casina Adams, Granddaughter of Point Hope, Alaska, Settler, Joseph Ferreira in 2005, Regional Oral History Office, Bancroft Library, Berkeley, University of California, 2007; Lowenstein, *Ultimate Americans*; Warrin, *So Ends This Day.*

20. Lisa Wexler, "The Importance of Identity, History, and Culture in the Wellbeing of Indigenous Youth," *Journal of the History of Childhood and Youth* 2, no. 2 (2009): 267–276.

21. A. C. Willox et al., "Examining Relationships between Climate Change and Mental Health in the Circumpolar North," *Regional Environmental Change* 15 (2015): 169–182.

22. Macolm McFee, "The 150% Man: A Product of Blackfeet Acculturation," *American Anthropologist* 70 (1968): 1096–1103.

23. Bodenhorn, "'He Used to Be My Relative'"; Mark S. Cassell, "Iñupiat Labor and Commercial Shore Whaling in Northern Alaska," *Pacific Northwest Quarterly* 91, no. 3 (2000): 115–123; Cassell, "Iñupiat Eskimo Workers in the Commercial Whaling Industry of North Alaska," *Arctic Studies Center Newsletter* 9 (September 2001): 14–15; Cassell, "The Landscape of Iñupiat Eskimo Industrial Labor," *Historical Archaeology* 39, no. 3 (2005): 132–151.

24. Catherine S. Reimer, "The Concept of Personal Well-being in the Iñupiat Worldview and Their View of Counselor Effectiveness" (PhD diss., George Washington University, 1996); Reimer, *Counseling the Iñupiat Eskimo* (Santa Barbara: Praeger, 1999); I. W. Borowsky et al., "Suicide Attempts Among American Indian and Alaska Native Youth: Risk and Prospective Factors," *Pediatrics and Adolescent Medicine* 153 (1999): 573–580; L. B. Whitbeck et al., "Discrimination, Historical Loss and Enculturation: Culturally Specific Risk and Resiliency Factors for Alcohol Abuse among American Indians," *Journal of Studies on Alcohol* 65 (2004): 409–418; J. White and N. Jodain, *Aboriginal Youth: A Manual of Promising Suicide Prevention Strategies* (Calgary: Center for Suicide Prevention, 2004); Wexler, "Importance of Identity, History, and Culture in the Well-being of Indigenous Youth."

25. Keith Basso, *Wisdom Sits in Places: Landscape and Language among the Western Apache* (Albuquerque: University of New Mexico Press, 1996); Yi-Fu Tuan, *Cosmos and Hearth: A Cosmopolite's View Point* (Minneapolis: University of Minnesota Press, 1999); Chie Sakakibara, "Our Home Is Drowning: Climate Change and Iñupiat Storytelling in Point Hope, Alaska," *Geographical Review* 98, no. 4 (2008): 456–475; Bruce Evan Goldstein, "Introduction: Crisis and Collaborative Resilience," in *Collaborative Resilience: Moving Through Crisis to Opportunity*, ed. B. E. Goldstein (Cambridge, MA: The MIT Press, 2012).

26. Lisa Wexler et al., "Central Role of Relatedness in Alaska Native Youth Resilience: Preliminary Themes from One Site of the Circumpolar Indigenous Pathways to Adulthood (CIPA) Study," *American Journal of Community Psychology* 52 (2013): 393–405; Wexler, "Looking Across Three Generations of Alaska Natives to Explore How Culture Fosters Indigenous Resilience," *Transcultural Psychiatry* 51, no. 1 (2014): 73–92.

27. Norman Chase, *The Iñupiat and Arctic Alaska: An Ethnography of Development* (San Diego: Harcourt, 1997); Cassell, "Iñupiat Labor and Commercial Shore Whaling."

28. Joseph G. Jorgensen, *Oil Age Eskimos* (Berkeley: University of California Press, 1990), xiv.

29. Ibid., 6.

30. Barbara Bodenhorn, "It's Traditional to Change: A Case Study of Strategic Decision-making," *Cambridge Anthropology* 22, no. 1 (2001): 24–51.

31. Ann Fienup-Riordan, *Eskimo Essays: Yup'ik Lives and How We See Them* (New Brunswick: Rutgers University Press, 1990).

32. N. Scott Momaday, "Native American Attitudes to the Environment," in *Seeing with a Native Eye: Essays on Native American Religion*, ed. W. Capp (New York: Harper and Row, 1974); Richard K. Nelson, *Shadow of the Hunter: Stories of Eskimo Life* (Chicago: University of Chicago Press, 1980); Basso, *Wisdom Sits in Places.*

33. Nelson, *Shadow of the Hunter*; Fienup-Riordan, *Eskimo Essays*; Karen Brewster, *The Whales, They Give Themselves: Conversations with Harry Brower, Sr.* (Fairbanks: University of Alaska Press, 2004).

34. E. Duran et al., "Healing the American Indian Soul Wound," in *International Handbook of Multigenerational Legacies of Trauma*, ed. Y. Danieli (New York: Springer, 1998); Wexler, "Importance of Identity, History, and Culture in the Wellbeing of Indigenous Youth."

35. L. Jilek-Aall, "Suicidal Behavior Among Youth: A Cross-Cultural Comparison," *Transcultural Psychiatric Research Review* 25 (1988): 87–105; Wexler, "Importance of Identity, History, and Culture in the Wellbeing of Indigenous Youth."

36. Sakakibara, "Our Home Is Drowning"; Sakakibara, "No Whale, No Music."

37. Sakakibara, "No Whale, No Music."

38. Aaron A. Fox, "Repatriation as Reanimation Through Reciprocity," in *The Cambridge History of World Music: Vol 1. (North America)*, ed. P. Bohlman (Cambridge: Cambridge University Press, 2014), 544.

39. Sakakibara, "Our Home Is Drowning."

40. Jean L. Briggs, "From Trait to Emblem and Back: Living and Representing Culture in Everyday Inuit Life," *Arctic Anthropology* 34 (1997): 227–235.

41. R. Linton, "Nativistic Movements," *American Anthropologist* 45 (1943): 230–240; R. Wagner, *The Invention of Culture* (Chicago: University of Chicago Press, 1975); Briggs, "From Trait to Emblem and Back."

42. Briggs, "From Trait to Emblem and Back"; Louis-Jacques Dorais, "Inuit Identity in Canada," *Folk* 30 (1988): 23–31.

43. Nobuhiro Kishigami, "Cultural and Ethnic Identities of Inuit in Canada," *Senri Ethnological Studies: Circumpolar Ethnicity and Identity* 66 (2004): 81–93.

44. Yi Wang, "Globalization Enhances Cultural Identity," *Intercultural Communication Studies* XVI, no. 1 (2007): 83–86.

45. Cassell, "Iñupiat Labor and Commercial Shore Whaling."

46. Lowenstein, *Ultimate Americans*; Warrin, *So Ends This Day*.

47. Rosita Worl, "The North Slope Iñupiat Whaling Complex," in *Alaska Native Culture and History*, ed. Y. Kotani and W. Workman, Senri Ethnological Studies 4

(Osaka: National Museum of Ethnology, 1980); David Boeri, *People of the Ice Whale: Eskimos, White Men, and the Whale* (New York: Dutton, 1983); R. L. Zumwalt, "Return of the Whale: Nalukataq, the Point Hope Whale Festival," in *Time Out of Time: Essays on the Festival*, ed. A. Falassi (Albuquerque: University of New Mexico Press, 1988); Edith Turner, "The Whale Decides: Eskimos' and Ethnographer's Shared Consciousness in the Ice," *Études/Inuit/Studies* 14 (1990): 39–42; Tom Lowenstein, *The Things That Were Said of Them: Shaman Stories and Oral Histories of the Tikiġaq People* (Berkeley: University of California Press, 1992); Lowenstein, *Ancient Land: Sacred Whale: The Inuit Hunt and Its Rituals* (New York: Farrar, Straus and Giroux, 1993); Bill Hess, *Gift of the Whale: The Iñupiat Bowhead Hunt, A Sacred Tradition* (Seattle: Sasquatch Books, 1999); Brewster, *The Whales, They Give Themselves*; Sakakibara, "Our Home Is Drowning"; Sakakibara, "No Whale, No Music"; Sakakibara, "Kiavallakkikput Aġviq (Into the Whaling Cycle): Cetaceousness and Global Warming among the Iñupiat of Arctic Alaska," *Annals of the Association of the American Geographers* 100, no. 4 (2010): 1003–1012.

48. Lowenstein *The Things That Were Said of Them*; Lowenstein, *Ancient Land.*

49. Lowenstein, *The Things That Were Said of Them*; Reimer, "The Concept of Personal Well-being in the Iñupiat Worldview"; Reimer, *Counseling the Iñupiat Eskimo.*

50. Chie Sakakibara, "Whale Tales: People of the Whales and Climate Change in the Azores," *FOCUS on Geography* 54, no. 3 (2011): 75–90.

51. F. Terra, *A Caça à Baleia nos Açoreano* (Ponta Delgada, Açores: Signo, 1949); Terra, *Natal Açoreano* (Ponta Delgada, Açores: Signo, 1958): D. de Melo, *Pedras Negras* (Lisbon: Vega, 1964); R. Guerra, *Trutas* (Horta: Câmara Municipal e Direcção Regional dos Assuntos Culturais, 1988); C. C. Emilio, *Azorean Folk Customs* (San Diego: Portuguese Historical Center, 1990); S. G. Costa, *Azores: Nine Islands, One History* (Berkeley, CA: Institute of Governmental Studies Press, 2008).

52. Onesimo Almeida, *Açores, Açorianos, Açorianidade* (Ponta Delgada, Açores: Signo, 1989).

53. Emilio, *Azorean Folk Customs.*

54. J. M. Figueiredo, *Introdução ao Estudo da Indústria Baleeira Insular* (Pico, The Azores, Portugal: Museu dos Baleeiros, 1996).

55. F. D. Santos et al., "Climate Change Scenarios in the Azores and Madeira Islands," *World Resource Review* (2004): 473–491.

56. Ibid.

57. A. Klimt and J. Leal, "Introduction: The Politics of Folk Culture in the Lusophone World," *Etnográfica* 9, no. 1 (2005): 5–17.

58. Ashley Lopes, "Teachers Become Students of Eskimo Culture," *South Coast Today*, August 8, 2004, http://www.southcoasttoday.com/article/20040808/news/308089984.

59. Emanuel J. Drechsel and T. Haunani Makuakāne, "Hawaiian Loanwords in Two Native American Pidgins," *International Journal of American Linguistics* 48, no. 4 (1982): 460–467.

Bibliography

Almeida, Onesimo T. *Açores, Açorianos, Açorianidade*. Ponta Delgada, Açores: Signo, 1989.

Basso, Keith. *Wisdom Sits in Places: Landscape and Language among the Western Apache*. Albuquerque: University of New Mexico Press, 1996.

Bockstoce, John. "Eskimo Whaling in Alaska." *Alaska Magazine* 43, no. 9 (1977.): n.p.

———. *Steam Whaling in the Western Arctic*. New Bedford, MA: Old Dartmouth Historical Society, 1977.

———. "History of Commercial Whaling in Arctic Alaska." *Alaska Geographic* 5, no. 4 (1978): n.p.

Bodenhorn, Barbara. "'He Used to Be My Relative': Exploring the Bases of Relatedness Among Iñupiat of Northern Alaska." In *Cultures of Relatedness: New Approach to the Study of Kinship*, edited by J. Carsten, 128–148. Cambridge: University of Cambridge Press, 2000.

———. "It's Traditional to Change: A Case Study of Strategic Decision-making." *Cambridge Anthropology* 22, no. 1 (2001): 24–51.

———. "Is Being 'Iñupiaq' a Form of Cultural Property?" In *Properties of Culture—Culture as Property, Pathways to Reform in Post-Soviet Siberia*, edited by E. Kasten, 35–50. Berlin: Dietrich Reimer Verlag, 2004.

Boeri, David. *People of the Ice Whale: Eskimos, White Men, and the Whale*. New York: Dutton, 1983.

Borowsky, I. W., M. D. Resnick, M. Ireland, and R. W. Blum. "Suicide Attempts Among American Indian and Alaska Native Youth: Risk and Prospective Factors." *Pediatrics and Adolescent Medicine* 153 (1999): 573–580.

Brewster, Karen. *The Whales, They Give Themselves: Conversations with Harry Brower, Sr.* Fairbanks: University of Alaska Press, 2004.

Briggs, Jean L. "From Trait to Emblem and Back: Living and Representing Culture in Everyday Inuit Life." *Arctic Anthropology* 34 (1997): 227–235.

Brower, Charles. *The Northernmost American: An Autobiography*. Typescript (895 pp.). Dartmouth College Library Stefansson Collection, n.d.

Burch, Ernest S. *The Traditional Eskimo Hunters of Point Hope, Alaska: 1800–1875*. Barrow: North Slope Borough, 1981.

Cassell, Mark S. "Iñupiat Labor and Commercial Shore Whaling in Northern Alaska." *The Pacific Northwest Quarterly* 91, no. 3(2000): 115–123.

———. "Iñupiat Eskimo Workers in the Commercial Whaling Industry of North Alaska." *Arctic Studies Center Newsletter* 9 (September 2001): 14–15. Washington, DC: Smithsonian Institution and National Museum of Natural History.

———. "The Landscape of Iñupiat Eskimo Industrial Labor." *Historical Archaeology* 39, no. 3 (2005): 132–151.

Chase, Norman A. *The Iñupiat and Arctic Alaska: An Ethnography of Development*. San Diego, CA: Hartcourt, 1990.

Costa, S. G. *Azores: Nine Islands, One History*. Berkeley, CA: Institute of Governmental Studies Press, 2008.

Deloria, E. *Speaking of Indians*. New York: Friendship Press, 1944.

Demallie, R. J. "Kinship and Biology in Sioux Culture." In *North American Indian Anthropology: Essays on Society and Culture*, edited by Raymond J. DeMallie and Alfonso Ortiz, 125–146. Norman: University of Oklahoma Press, 1994.

de Melo, D. *Pedras Negras*. Lisbon: Vega, 1964.

Drechsel, Emanuel J., and T. Haunani Makuakāne. "Hawaiian Loanwords in Two Native American Pidgins." *International Journal of American Linguistics* 48, no. 4 (1982): 460–467.

Dorais, L.-J. "Inuit Identity in Canada." *Folk* 30 (1988): 23–31.

Duran, E., B. Duran, M. Yellow Horse-Davis, and S. Yellow Horse-Davis. "Healing the American Indian Soul Wound." In *International Handbook of Multigenerational Legacies of Trauma*, edited by Y. Danieli, 341–354. New York: Springer, 1998.

Emilio, C. C. *Azorean Folk Customs*. San Diego, CA: Portuguese Historical Center, 1990.

Figueiredo, J. M. *Introdução ao Estudo da Indústria Baleeira Insular*. Pico, The Azores, Portugal: Museu dos Baleeiros, 1996.

Fienup-Riordan, Ann. *Eskimo Essays: Yup'ik Lives and How We See Them*. New Brunswick: Rutgers University Press, 1990.

Fox, Aaron A. "Repatriation as Reanimation Through Reciprocity." In *The Cambridge History of World Music: Vol. 1 (North America)*, edited by P. Bohlman, 522–554. Cambridge: Cambridge University Press, 2014.

Goldstein, Bruce Evan. "Introduction: Crisis and Collaborative Resilience." In *Collaborative Resilience: Moving Through Crisis to Opportunity*, edited by B. E. Goldstein, 1–15. Cambridge, MA: The MIT Press, 2012.

Guerra, R. *Trutas*. Horta: Câmara Municipal e Direcção Regional dos Assuntos Culturais, 1988.

Harkin, M. E. "Ethnographic Deep Play: Boas, McIlwraith, and Fictive Adoption on the Northwest Coast." In *Strangers to Relatives: The Adoption and Naming of Anthropologists in Native North America*, edited by Sergei Kan. Lincoln: University of Nebraska Press, 2001.

Hess, Bill. *Gift of the Whale: The Iñupiat Bowhead Hunt, A Sacred Tradition*. Seattle: Sasquatch Books, 1999.

Jilek-Aall, L. "Suicidal Behavior Among Youth: A Cross-Cultural Comparison." *Transcultural Psychiatric Research Review* 25 (1988): 87–105.

Jorgensen, Joseph G. *Oil Age Eskimos*. Berkeley: University of California Press, 1990.

Kan, Sergei. "Friendship, Family, and Fieldwork: One Anthropologist's Adoption by Two Tlingit Families." In *Strangers to Relatives: The Adoption and Naming of Anthropologists in Native North America*, 185–217. Lincoln: University of Nebraska Press, 2001.

Kishigami, Nobuhiro. "Cultural and Ethnic Identities of Inuit in Canada." *Senri Ethnological Studies: Circumpolar Ethnicity and Identity* 66 (2004): 81–93.

Klimt, A., and Leal, J. "Introduction: The Politics of Folk Culture in the Lusophone World." *Etnográfica* 9, no. 1 (2005): 5–17.

Linton, R. "Nativistic Movements." *American Anthropologist* 45 (1943): 230–240.

Lopes, Ashley. "Teachers Become Students of Eskimo Culture." *South Coast Today*, August 8, 2004, http://www.southcoasttoday.com/article/20040808/news/308089984.

Lowenstein, Tom. *The Things That Were Said of Them: Shaman Stories and Oral Histories of the Tikiġaq People*. Berkeley: University of California Press, 1992.

———. *Ancient Land: Sacred Whale: The Inuit Hunt and Its Rituals*. New York: Farrar, Straus and Giroux, 1993.

———. *Ultimate Americans: Point Hope, Alaska: 1826–1909*. Fairbanks: University of Alaska Press, 2008.

McFee, Malcolm. "The 150% Man: A Product of Blackfeet Acculturation." *American Anthropologist* 70 (1968): 1096–1103.

Melville, Herman. *Moby-Dick; or, The Whale*. New York: Harper and Brothers, 1851.

Momaday, N. Scott. "Native American Attitudes to the Environment." In *Seeing with a Native Eye: Essays on Native American Religion*, 79–85, edited by W. Capp. New York: Harper and Row, 1974.

Nelson, Richard K. *Shadow of the Hunter: Stories of Eskimo Life*. Chicago: University of Chicago Press, 1980.

Noice, Harold. *With Stefansson in the Arctic*. London: George G. Harrap and Co., Ltd., 1924.

Offen, Karl. "Making Black Territories." In *Mapping Latin America: A Cartographic Reader*, edited by J. Dym and K. Offen, 288–292. Chicago: The University of Chicago Press, 2011.

Peluso, N. L. "Whose Woods Are These?: Counter-Mapping Forest Territories in Kalimantan, Indonesia." *Antipode* 27, no. 4 (1995): 383–406.

Quijano, A. "Colonality and Power, Eurocentrism, and Latin America." *Nepentla: Views from the South* 1, no. 3 (2000): 533–580.

Reimer, Catherine S. "The Concept of Personal Well-being in the Iñupiat Worldview and Their View of Counselor Effectiveness." Doctoral dissertation, George Washington University, 1996.

———. *Counseling the Iñupiat Eskimo*. Santa Barbara, CA: Praeger, 1999.

Sakakibara, Chie. "Our Home Is Drowning: Climate Change and Iñupiat Storytelling in Point Hope, Alaska." *Geographical Review* 98, no. 4 (2008): 456–475.

———. "No Whale, No Music: Contemporary Iñupiaq Drumming and Global Warming." *Polar Record* 45, no. 4 (2009): 289–303.

———. "Kiavallakkikput Aġviq (Into the Whaling Cycle): Cetaceousness and Global Warming among the Iñupiat of Arctic Alaska." *Annals of the Association of the American Geographers* 100, no. 4 (2010): 1003–1012.

———. "Climate Change and Cultural Survival in the Arctic: Muktuk Politics and the People of the Whales." *Weather, Climate and Society* 3, no. 2 (2011): 76–89.

———. "Whale Tales: People of the Whales and Climate Change in the Azores." *FOCUS on Geography* 54, no. 3 (2011): 75–90.

Santos, F. D., et al. "Climate Change Scenarios in the Azores and Madeira Islands." *World Resource Review* (2004): 473–491.

Shoemaker, Nancy. "Whale Meat in American History." *Environmental History* 10, no. 2 (2005): 269–294.

———. *Native American Whalemen and the World: Indigenous Encounters and the Contingency of Race*. Chapel Hill: University of North Carolina Press, 2014.

Spruhan, P. "Legal History of Blood Quantum in Federal Indian Law to 1936." *South Dakota Review* 51 (2006): 1–50.

TallBear, K. "Narratives of Race and Indigeneity in the Genographic Project." *Journal of Medicine Ethics and Law* 35 (2007): 412.

Terra, F. *A Caça à Baleia nos Açoreano*. Ponta Delgada, Açores: Signo, 1949.

———. *Natal Açoreano*. Ponta Delgada, Açores: Signo, 1958.

Tuan, Yi-Fu. *Cosmos and Hearth: A Cosmopolite's View Point*. Minneapolis: University of Minnesota Press, 1999.

Turner Edith. "The Whale Decides: Eskimos' and Ethnographer's Shared Consciousness in the Ice." *Études/Inuit/Studies* 14 (1990): 39–42.

Wagner, R. *The Invention of Culture*. Chicago: University of Chicago Press, 1975.

Wang, Yi. "Globalization Enhances Cultural Identity." *Intercultural Communication Studies* XVI, no. 1 (2007): 83–86.

Warrin, Don. Interview transcript by Don Warrin with Elsie Casina Adams, Granddaughter of Point Hope, Alaska, Settler, Joseph Ferreira in 2005. Regional Oral History Office, Bancroft Library. Berkeley, University of California, 2007.

———. *So Ends This Day: The Portuguese in American Whaling 1765–1927.* North Dartmouth: University of Massachusetts Dartmouth Center for Portuguese Studies and Culture, 2010.

Wexler, Lisa. "The Importance of Identity, History, and Culture in the Wellbeing of Indigenous Youth." *The Journal of the History of Childhood and Youth* 2, no. 2 (2009): 267–276.

———. "Looking Across Three Generations of Alaska Natives to Explore How Culture Fosters Indigenous Resilience." *Transcultural Psychiatry* 51, no. 1 (2014): 73–92.

Wexler, Lisa, Joshua Moses, Kim Hopper, Linda Joule, Joseph Garoutte, and the LSC CIPA Team. "Central Role of Relatedness in Alaska Native Youth Resilience: Preliminary Themes from One Site of the Circumpolar Indigenous Pathways to Adulthood (CIPA) Study." *American Journal of Community Psychology* 52 (2013): 393–405.

Whitbeck, L. B., X. Chen, D. R. Hoyt, and G. W. Adams. "Discrimination, Historical Loss and Enculturation: Culturally Specific Risk and Resiliency Factors for Alcohol Abuse among American Indians."*Journal of Studies on Alcohol* 65 (2004): 409–418.

White, J., and N. Jodain. *Aboriginal Youth: A Manual of Promising Suicide Prevention Strategies.* Calgary: Center for Suicide Prevention, 2004.

Whyte, K. P. "Indigeneity and US Settler Colonialism." In *Oxford Handbook of Philosophy and Race*, edited by Naomi Zack. Oxford: Oxford University Press, 2016.

Willox, A. C., E. Stephenson, J. Allen, F. Bourque, A. Drossos, S. Elgarøy, M. J. Kral, I. Mauro, J. Moses, T. Pearce, J. MacDonald, and L. Wexler. "Examining Relationships between Climate Change and Mental Health in the Circumpolar North." *Regional Environmental Change* 15 (2015): 169–182.

Wohlforth, C. *The Whale and the Supercomputer: On the Northern Front of Climate Change.* New York: North Point Press, 2005.

Worl, Rosita. "The North Slope Iñupiat Whaling Complex." In *Alaska Native Culture and History*, edited by Y. Kotani and W. Workman, 305–320. Senri Ethnological Studies 4. Osaka: National Museum of Ethnology, 1980.

Zumwalt, R. L. "Return of the Whale: Nalukataq, the Point Hope Whale Festival." In *Time Out of Time: Essays on the Festival,* edited by A. Falassi, 261–276. Albuquerque: University of New Mexico Press, 1988.

8

Landscapes on Hold

The Norwegian and Russian Barents Sea Coast in the New North

janike kampevold larsen & peter hemmersam /
OSLO SCHOOL OF ARCHITECTURE AND DESIGN

THIS ESSAY EXPLORES CHANGING IMAGES OF THE NORWEGIAN-Russian Barents coast landscape, which is currently undergoing political and economic changes due to large prospected quantities of extractable oil and gas, rare earth, and other minerals. It considers the relationship between the reductive projections that industries and governments lay on the remote territories of the North and the experience and agency of local communities that encounter new industries. This includes speculative responses to the possibility of future industrial exploitation. In other words, this chapter looks at the contrasting ways in which landscapes are socially and materially *produced* simultaneously by external expectations of future development and by local perceptual responses—including forms of activism, art, and architecture—to forces of globalization that render the future of many settlements uncertain.

We find ourselves in an age when the surface of the earth is more exposed and in a more precarious state than ever before, leading scientists to propose that we live in a new epoch, the Anthropocene. Humans have become accustomed to thinking of themselves as the prime agent in natural environments but are now increasingly becoming aware that other agencies are at play, including those of animals, nonhuman actors (as well as gods and traditional spiritual beliefs, as described in Carly Dokis's chapter in this volume, for example), geology, and the material properties of landscape itself. This reflects a new humility toward landscapes, one that reverts to Ian McHarg's mid-twentieth-century argument that human habitation should be "design[ed] with nature."[1] With reference to authors like Doreen Massey and David Harvey, the introduction to this volume argues that place is "dynamic, layered, connected, and socially constructed," and in line with contemporary landscape theory we adopt a similar perspective when it comes to landscapes.

Along the Barents coast, we find a number of towns and landscapes that are on hold; in different ways they are waiting for a future resolution to contemporary problems. These include Teriberka in Northwest Russia, which for a few years featured as the designated land hub for the Shtokman gas field in the Barents Sea (a plan that was put on hold in August 2012), and the Norwegian town of Vardø, which has experienced depopulation due to a decrease in the number of fishing licenses. Similarly, Hammerfest, a small town west of the northern tip of Norway, is currently prospering due to the liquid natural gas facilities that became operational in 2007 and serves as an example of a place that has profited from the oil industry—but its population is ironically more confused than in the prepetroleum cities of Teriberka and Vardø as it has to deal with urban change and shifting identities.

Vardø and Teriberka are both located on the Barents coast, roughly two hundred kilometers apart and separated by the Norwegian-Russian border. The two towns share a common history of commerce and fishing, and more recently a destiny of unemployment and depopulation. They differ greatly, however, when it comes to the political and economic contexts in which they exist. While Vardø holds importance as a site of military presence and cultural heritage in the Northeast of Norway, Teriberka functions as a satellite to the much larger and strategic city of Murmansk. Hammerfest is as old as Vardø and has a shared history of fishing and trade. All three have also functioned as bases for arctic exploration. Currently Hammerfest differs greatly from Vardø and Teriberka due to the income and increasing number of inhabitants that have accompanied the arrival of the petrochemical industry.

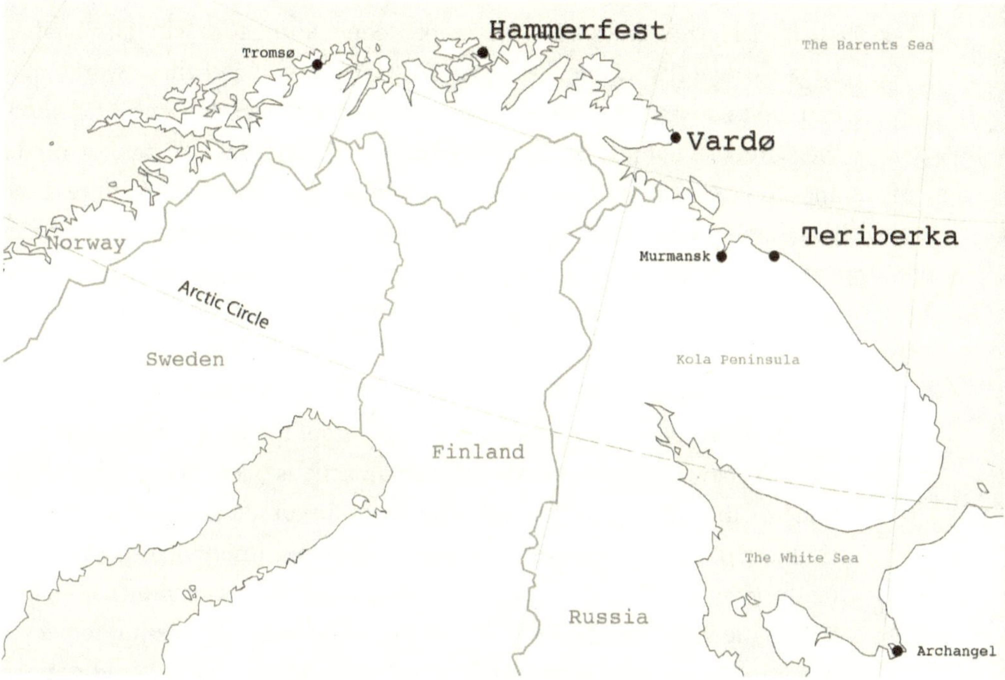

Fig. 8.1. Map of northern Scandinavia and the Barents region.

The three towns are in different ways landscapes "on hold"; they are caught between regional economic projections and the desires and expectation of local inhabitants in their everyday life as well as the materiality of the ground and the agency of nature. They are results of the outside projections of national and global interests that are predicting future possibilities and revenues. We have identified different ways in which the reading of, and intervention into, these landscapes may respond to these projections by being what American writer, art critic, and curator Lucy Lippard calls "'place specific' rather than 'site specific.'"[2] The thesis is that by insisting on the place-specific one expands the local capacity for thinking about landscape change. Lippard's point is that artists (as well as other agents dealing with visual culture) can play a role in expanding consciousness and even teach communities to see the emerging and changing landscapes by reading and addressing them as local places rather than remotely projected sites. This, Lippard argues, was the case with famous land-art artists like

Michael Heizer, Walter De Maria, or Robert Smithson, who all installed large earthworks in the American West. While some of these works may be read as abstractly critiquing nuclear testing (Heizer's *Double Negative*), the reductive context of the museum (Smithson's *Spiral Jetty*), or man's appropriation of land (De Maria's *Lightning Field*), they fail to address the vulnerable condition of living in these arid terrains, which have been speedily degraded by mining (uranium in particular), gravel industries, road building, and the increasingly congested transportation that serves these industries.

Current discourse on "the art of the Anthropocene," or simply Anthropocene art,[3] implies that it should adopt the long view and the big scale to show the consequences of human actions (akin to the cultural critique of the writing in the Anthropocene, discussed in Kyndra Turner's chapter). Lippard, however, believes that environmental impact due to (anthropogenic) climate change and extraction industries should compel art to become social and political in order to become locally relevant to people living in changing environments. If we follow Lippard, art should not just be site-specific and reference the physical context in which it is built—as has been fashionable since the 1960s—it should be place-specific; that is, it should include people and consider the network of social and environmental agencies in a given place. We argue that art (along with other types of local practices), by being increasingly relevant to communities, has a critical capacity to trigger energies in cities that are "on hold"—that is, in effect, a new level of politics in communities that are left "dangling" at the edge of national and international economic development and politics.

History

The Barents coast stretches from northern Norway into Northwest Russia, and due to the Gulf Stream its waters are largely ice-free in winter. It is a region that has historically had a small population and where national states and large-scale commercial interests were late in making their entrance.

Early historic settlements on the Barents coast were primarily Sami, and Sami communities still reside in the four national states of the region: Norway, Russia, Finland, and Sweden. National territorial claims in this region have historically been fluid. In the fourteenth century, a degree of stability was attained, delimitating Danish-Norwegian and Swedish interests against those of the Novgorod Republic, but for several hundred years—until 1602—Sami populations in certain areas were subject to taxation from both the Danish-Norwegian and the emerging Russian state. Pomor populations from the White Sea area and Karelian groups from the Baltic region gradually settled in what is today known as the Kola Peninsula in northwest Russia, and from the sixteenth century onward the Russian state gradually expanded its political and economic interests over the area. In the late nineteenth century, Nenets and Komi indigenous groups also migrated west into the Kola Peninsula.

The Norwegian settlement of the Barents coast began in the thirteenth century; in the nineteenth century Finnish groups also settled in what became known as Finnmark, particularly along the east coast of the Varanger Peninsula. The late nineteenth and early twentieth century saw the height of the Pomor trade along the Barents coast (timber, fish, and fur from Russia traded against a variety of local products, fish and fish oil being the most important ones), which tied the settlements along the coast together.

During both world wars, the Barents coast gained strategic importance as a region rich in minerals and as offering Russia ice-free access to the Atlantic. In particular, the Russian Kola Peninsula saw heavy industrialization and militarization in the period between and after the two wars. Large numbers of workers were imported, swelling the population of the peninsula to well over one million (compared to the seventy-five thousand inhabitants currently residing in the Norwegian border county of Finnmark). Sweden and the new twentieth-century Finnish state also had territorial interests in the Barents area, and for a period in the early twentieth century Finland controlled a land corridor to the Barents Sea—lost to Russia during World War II.

In the twentieth century a radical division replaced the once-fluid populations and geographies of this coastline. The Russian Revolution saw the closing of transborder trade and immigration, and social and economic development of the twentieth century differed radically on each side of the militarized border. Both Norway and the USSR, however, saw a strategic interest in sustaining and even expanding the populations on their respective sides of the border.

While today's cross-border trade remains limited, the movement of people has increased significantly over the last decades. In 2010 the remaining maritime border dispute between Russia and Norway was settled, and today the Barents coast is marked by a new focus on resource extraction (oil, gas, and minerals), the promise of new business associated with the opening of the Northern Sea Route (the Northeast Passage), as well as traditional sources of income such as the still-rich fishing and animal husbandry practices. The population along the Barents coast has decreased over the last decades, but the promise of a new regional economy is generating a new optimism.

Teriberka

The town of Teriberka, northeast of Murmansk, currently boasts a population of fourteen hundred, while it held up to twelve thousand inhabitants in the 1950s and '60s.[4] The economy was historically based on fishing and on trade with the White Sea area, northern Norway, and Europe. In the late nineteenth century, the Barents coast saw immigration from Finland and Norway encouraged by the Russian state. In the early twentieth century, a Norwegian-owned fish-processing facility and general merchandise store was established in Teriberka. In its heyday in the middle of the twentieth century, Teriberka's businesses also included mink farming, shark fishing, dairy farming, and reindeer herding.

The structural changes of the 1990s caused a collapse in Teriberka's economy. Unemployment rose and population numbers dropped. In 2007 the town was selected to become the Shtokman gas field development land hub, which would have required a temporary population of ten thousand during the construction of a new harbor, infrastructure, and a processing plant, with an expected six hundred permanent jobs when the gas field went into production.[5] In preparation for the Shtokman development, roads and other infrastructure were built and the social infrastructure was somewhat upgraded (Gasprom sponsored a kindergarten in the town). In 2012, however, the Shtokman project was put on hold, to the frustration of locals and private investors. The same year saw the opening of a modest

Fig. 8.2. The new road to the planned shipping facility passes by the cemetery in Teriberka—in preparation for a new industrial era for the town. PHOTO: P. HEMMERSAM (2012)

fish-processing plant, but today the town is again largely left at the margins of the driving economic forces on the Kola Peninsula: mining, military activities, and surveillance operations.

The town is in a state of dilapidation, with a few highly visible exceptions due to the activities of the local Shtokmann development office, and the Russian authorities demonstrate little active interest in cultural historic heritage or preservation of the many historic wooden structures in dire need of care. Empty buildings, abandoned wharves, and wooden boats rest in an entropic process of decomposition.

Vardø

Vardø was founded as a fortification around 1300 to support the Danish-Norwegian territorial claim in the region. It was a tax collection center for centuries and a market place for the Pomor trade. It is located on two conjoined islands off the coast on the extreme eastern tip of mainland Norway, a situation that even today makes it a prime location for monitoring Russian military activities, transpolar ballistic missile trajectories between Asia and North America, as well as shipping and offshore installations in the Barents Sea. There is a United States–Norwegian radar facility on top of the island.

Unlike most other towns of northern Norway, Vardø was only partially destroyed in World War II, leaving it with a number of buildings from the mid-nineteenth and early twentieth centuries. In 1968 the town had 4,222 inhabitants, but the current population is only 2,128.[6] The once-dominant fishing industry has declined dramatically: only two of fourteen fishing factories are left, and there are currently around seventy operating fishing boats in the municipality of Vardø, compared to the hundreds that were operational only twenty years ago, and the almost two thousand that would be assembled there for the winter fishing season. A coastal steamer, operated by Hurtigruten, supports a very limited income from tourism in the town.

Vardø has attempted to ready itself in the expectation of offshore activities (the town actually being closer to the Russian Shtokman gas fields than Teriberka is). A large new harbor has been constructed on the mainland, but it lies largely vacant—save for a king crab–processing facility—waiting for possible future petroleum-related business or Northern Sea Route supply service activities. Currently Vardø receives very little direct income from these new regional industries, but its stock of preserved wooden structures, its compelling charm, and its strategic position in relation to the Barents Sea oil reserves, as well as the Northern Sea Route, has led to detrimental property speculation. The decreasing population has left a number of buildings empty, and investors and property owners simply leave them to decay and ultimately demolition in order to make way for new development in case the economy of the town picks up and property values increase. The many vacant buildings of the town reinforce the perception of the town as a landscape on hold.

Local Practices

There are, however, actors that are not waiting for outside economic development to benefit the town. They enact different practices and are building a variety of businesses in order to develop Vardø in a sustainable fashion through a diversity of activities. One important initiative is represented by a group of local enthusiasts, led by Svein Harald Holmen, who has been working consistently over the last five years to both preserve and develop the town's architectural heritage. Holmen is now developing Vardø Restored, a platform aimed at inspiring local owners to invest in their city.[7] Many of Vardø's pre–World War II wooden buildings are also in dire need of renovation. In 2010 Holmen initiated a strategy to secure a large body of decaying wooden structures in and around Vardø, and funding from the Directorate for Cultural Heritage and the Norwegian Cultural Heritage Fund has allowed volunteers to partially preserve a few of the most acutely decaying buildings.[8] These funds are tied to refurbishing individual buildings and specific parts of buildings—windows, walls, foundations, etc.—but are important contributions to a task that ultimately requires much larger private investments. To a limited degree, this has spurred the owners of individual buildings to start investing in their properties, and in the case of one of the historic fish factories, investors have been convinced to preserve and restore existing buildings rather than wait for an uncertain future rise in property value.

The approach of Holmen and his fellow local activists includes very deliberate attempts to involve the local community in deciding the future uses of old buildings, both to secure investment in their refurbishment and also to build a shared sense of ownership in and obligation to the town. Identifying and visualizing the problem and potential has been a critical approach of the group, and in one particular instance the town and its buildings were mapped—and put onto the map—through street art.

Street Art

In the summer of 2012 the town of Vardø commissioned the Norwegian artist Pøbel to install street art. He in turn invited twelve European street artists, who, over the course of a few brief summer weeks, executed fifty-four pieces on buildings and other structures—either with textual slogans and cartoonlike drawings or in a more figurative style. Obtaining permission from all building owners, the project served as a marking of

Fig. 8.3. Street art by E. B. Itso in Vardø from the Koma Festival in 2012. Quote from Herman Melville's *Moby-Dick*. Several old abandoned buildings became canvases for the visiting artists. PHOTO: P. HEMMERSAM (2013)

precariously dilapidated buildings. The project helped to turn attention not only to the endangered building mass but also to the dramatic decline in population. The art project became an artistic tool in the development of a local community facing social and economic problems. It unfolded as a place-specific map of the town; it involved local property owners and affected public perception of the state of the town. In the aftermath individuals who wanted to preserve and restore them in fact bought some of the buildings involved.

Birding and Architecture

Another place-specific practice in Vardø is the architectural practice Biotope,[9] a collaborative that has found a new and alternative audience in the worldwide community of bird-watchers to whom Vardø is a unique location. With the support of regional authorities, its members have carried out a feasibility study of the development of Vardø and the Varanger Peninsula as a birding destination in which they mapped the natural, economic, and cultural resources of the area.[10] In a number of locations in and around Vardø they have designed and built shelters for bird-watching, partnered with and encouraged local entrepreneurs, worked with place branding and the media, and organized birding festivals that brought visitors from around the world. In addition, they have helped the local community understand and appreciate its own (unique) landscape setting. As such, Biotope is an example of the active engagement with the agency of landscape in local and regional development, through the medium of design and architecture. Its architectural landscape interventions take the form of delicately designed pavilions for temporary landscape inhabitation that are located in the landscape according to an intimate understanding of bird behavior (and a pragmatic knowledge of the preferences of bird-watchers). These shelters are not just *site-specific* enhancements of landscape vistas (such as the follies of the classic romantic garden) but represent a *place-specific* and deliberate

Fig. 8.4. The bird-watching shelter on Steilnes by Biotope is carefully sited at a meaningful location. PHOTO: BIOTOPE/TORMOD ANDERSEN

underlying engagement with the organic inhabitation of the landscape—through enhanced vision technologies in the form of powerful telephoto lenses and binoculars. The project facilitates for a global community of bird-watchers and indirectly addresses the dire need for new income and jobs on the Varanger Peninsula, despite a lack of explicit political and economical perspective in Biotope's mission.

Hammerfest

Place-specific aesthetic practices as a tool of awareness-raising have also been employed in Hammerfest. In May 2013 Hanne Hammer Stien, the curator of the national art festival, invited several Norwegian artists to create and install their art in various locations of their own choice in the town. The architectural/artistic practice *Fellesskapsprosjektet å Fortette Byen* (FFB) (The Community Project to Densify the City) installed themselves on a steep escarpment at the back of this windblown arctic town, in a small cave that was excavated during World War II.[11] They built a fragment of infrastructure from recycled material: a monorail up to the cave. Inside the cave FFB installed a video featuring the moving of a stack of discarded timber—a pile that had gradually accumulated during the construction of the liquid natural gas plant. This logistical ballet involved a coastal freighter—costing $200,000—and thirty trucks shuttling back and forth for a day to bring five thousand tons of leftover material from the pile to the freighter and on to Sweden where it would be incinerated.

In front of the cave, on a narrow plateau, FFB established a small garden. By building a fireplace, serving food, and hosting midnight sun parties every night of the festival, they appropriated a *space* and made it a *place*. This happened by transforming the material making of the site and involving visitors in social acts, thereby changing their perception of it and giving it new meaning. With artists Kristin Tårnes and Margrethe Pettersen they served a Russian-style borscht soup, brewed over the Tromsø palm—and thereby moved beyond the quite worn topos of ad hoc art performances by also providing food. The Tromsø palm (*Heracleum persicum*)[12] is an invasive species of the carrot family that actively kills off neighboring plants, leaving it dominant in large areas.

Fig. 8.5. FFB's installation in the cave above Hammerfest was a social space for a brief period. PHOTO: FFB

The project as a whole addressed sustainability by using low-cost recycled materials in a town otherwise marked by industrial extravagance, but it also addressed the appropriation of land by extractive industries. The frail infrastructure leading up to the cave, as well as the invitation to the festival audience to populate the plateau in front of it, overlooking the town and the liquid natural gas plant at Melkøya, served as a multilayered comment on the industrial appropriation of arctic and subarctic lands and territories. These industries generally appropriate land by an infrastructure that does not necessarily lead to an increase in population, only to increased revenue for investors. FFB's project emblematically demonstrates exactly how the site-specific seems to have become obsolete in regions under pressure. Like the American West—which Lippard describes in her most recent book *Undermining* as becoming "heavily dependent on tourism, along with military, nuclear, and extractive industries"[13]—the Norwegian North is becoming object to intensified extraction, its landscapes subject to intensified commodification, and its populations exposed to a very rapid revelation of new opportunities, which is followed by disorientation. Surprisingly, FFB's project makes a poetic shift toward contemplating the values of the everyday vernacular landscape, its local resources, its close and distant views, and, most important, its sociability. While addressing the forces at work in the large subarctic region, their work is place-specific.

Thinking Places

Place thinking in art and architecture is closely related with thinking about, and intervening in, landscape. Lucy Lippard has been writing about global land art since its very beginning in the 1960s, when Robert Smithson, Michael Heizer, Walter De Maria, and others installed different and often quite abstract works of art in the extreme landscapes of the American West. At that time a New West was also emerging due to the industrial boom in the 1960s, forcing a radical rethinking of the region. In a recent essay Lippard states that while the classic land art provided an important visual expansion of consciousness (breaking the confines of the gallery), contemporary site-specific art can no longer rely on the same recipe and seems to a lesser degree capable of expanding consciousness and absorbing and reflecting regional (cultural, economic, and climate) change:

> Monumental land art takes too much of its power from distance—distance from people, from places, and from issues—while my own interest had come to focus almost entirely on the nearby, on specific places as they reflect the interactions between people and what we call "nature," which *includes* people.[14]

While being social in the sense that it caused young artists to physically travel to a remote and socially different area of the country, the classic land art movement maintained a distance from the land artists built in that implied a rejection of social engagement. Arguing for a "microview of land and art, and for grassroots connections,"[15] Lippard insists that artists should take responsibility for the landscape they install themselves in—by becoming intimate not only with the landscape but also with the social and political context of a *place* in order to develop "human responses to a specific landscape."[16]

Lippard claims that climate change, decreasing resources (which is one of the key arguments for arctic oil drilling), and industrial management that sacrifices the environment for economic gains (for example, oil sands) force artists out of their usual habitats. They need to take responsibility for what she calls "confused populations"—populations that are uncertain about how to choose to live in monoindustrial settlements. She thus echoes a claim by Frans Berkhout that "the Anthropocene analysis, with its focus on the global and the long run . . . appears to leave the predicaments faced by people at a distance."[17] While Berkhout argues that the Anthropocene needs to be taken down to "spatial and social scales that matter to people," Lippard argues that art that addresses the Anthropocene needs to widen its scope in a way that includes the social and political texture of places that involve people. Artistic practices have to move beyond being *site-specific* (an art-scene catchphrase of the last couple of decades) to become *place-specific*, and to consider the network of social, economic, and environmental agencies in any given place. This might be what William L. Fox describes as the third stage of Anthropocene art—where the artist actually intervenes in the systems of the earth.[18]

Thinking Futures

Hammerfest, Vardø, and Teriberka reside within one of the last territories to be exploited on our planet—one that is exposed to extranational and multicorporate interests on a scale and at a speed that are unprecedented. These towns along the Barents coast exist in a radical tension between their local, geologic, and microclimatic situations and global economic interests. They can be seen as landscapes held in stasis: physical and perceived landscapes that are the temporary results of projected future economic and social changes. They are the products of the politics of globalization but also of the desires and practices of local actors and inhabitants. And in them, local community and landscape agency is acted out in very different ways. These towns demonstrate, in three different stages, that the *oil-specific* city is in no way *site-specific*—they subsist in a network of global and national economic, as well as strategic, interests. Teriberka is left with an upgraded kindergarten, a large new road rushing by its cemetery, and broken hopes of being part of the new international petroleum economy while fishing boats are rotting on the beaches. Here, development stalled in the course of a few days in 2012, and the place was abandoned to its former state of dilapidation with no prospective means of sustenance. Vardø is being rebuilt to meet petroleum development while also expecting to survive on tourism. The protected and picturesque inner harbor between the two inhabited islands is being supplanted by a large new harbor on the mainland, the future use of which remains uncertain. Hammerfest is in the later stage of already having built expensive cultural infrastructure financed by real estate tax income from the petroleum industry but—just as the other towns—finds itself in a situation in which the future has not been sustainably resolved.

These towns are understood, both by media and also by local economic and administrative forces, to be subject to projections of the future from the outside, often based on statistics and numbers.[19] Looking closely, however, what appears to be homogenizing external force actually produces difference locally—but it also produces a general uncertainty in facing the future, resulting in landscapes that are put on hold, waiting for something to happen.

In future developments along the Barents coast—or indeed anywhere in the changing Arctic—genuine local stakeholder involvement is a prerequisite to sustainable future thinking. Towns like Teriberka, Vardø, and Hammerfest are in need of a locally based reading of agency—as a way to unravel or nuance the hegemonic understanding of transformations in the Arctic that forms them at the same time as it ignores local conditions.

Place-specificity does, however, hold its own problems. For instance, emphasizing the importance of place thinking in aesthetic practices may in certain instances lead to a nostalgia for a more immediate or "authentic" relationship between people and space, which could potentially blind one to the double dynamics of place production in capitalism[20]: the "abstraction of space and the 'production' of particularities of place, local specificity and cultural authenticity."[21] Capitalism itself seeks difference and authenticity and "our willingness to pay high prices for it only highlights the degree to which they are already lost to us (thus the power they have over us)."[22] This dynamic is not yet prominent in the landscapes that we have looked into.

Just as the Old West was transforming into a New West at the time the classic land art movement emerged, the North is now rapidly transforming into a New North that requires an expansion of consciousness to enable future thinking. In order to escape the preconfigured risk of acting as designers/researchers[23] who anticipate and project futures for the Arctic, one has to work with the local communities to enhance, through critical analysis, the geographical imaginary involved in the development of the New North. This is the methodology of what we call the Future North, which is more than site-specific but also place-specific, sited both in space and time, and not only responding to projected scenarios but recognizing a multitude of local agencies, including that of local community and also that of landscape itself.

This approach seeks to avoid the topoi of the Arctic as desolate, uncharted, and even uninhabited that still dominate policies on mining and other forms of resource extraction. For most of the world, the Arctic is a repository of resources (industry), data (science), and experiences (tourism), and at the same time its landscapes are seen as (passive) receivers of pollution, infrastructure, and people. The resource industry, strategic policy makers, the tourism industry, and even researchers seeking local and authentic perspectives on the Arctic all impose different projections of what a place in the Arctic is and can be. These are, however, seldom based on a dialogical association with the places themselves—drawn as they are between external projections, strong forces of globalization, and local social dynamics. By engaging in this methodology, we seek to bring attention to both practices that emerge from within these landscapes and ways in which communities may have independent and connected agency in developing visions for the future of the North.

Endnotes

1. Ian McHarg, *Design with Nature* (New York: Doubleday, 1971).

2. Lucy Lippard, "Peripheral Vision," in *Land Arts of the American West*, ed. Chris Taylor and Bill Gilbert (Austin: University of Texas Press, 2009), 338.

3. William L. Fox, "The Art of the Anthropocene," in *Curating the Future*, ed. Jennifer Newell, Libby Robin, and Kirsten Wehner (Honolulu: University of Hawai'i Press, 2016); Libby Robin, Dag Avango, Luke Keogh, Nina Möllers, Bernd Scherer, and Helmuth Trischler, "Three Galleries of the Anthropocene," *The Anthropocene Review* (September 2014); Elizabeth Ellsworth and Jamie Kruse, ed., *Making the*

Geologic Now: Responses to Material Conditions of Contemporary Life (New York: Punctum Books, 2013).

4. Nils Aarsæther, Larissa Riabova, and Jørgen Ole Bærenholdt, "Community Viability," in *Arctic Human Development Report*, ed. Nils Einarsson, Joan Nymand Larsen, Annika Nilsson, and Oran R. Young (Akureyri: Stefansson Arctic Institute, 2004).

5. "The Shtokman Office in Teriberka," *Barents Observer*, June 9, 2009, http://barentsobserver.com/en/node/18632.

6. Statistics Norway, 2013.

7. See http://vardorestored.com.

8. In 2013 the group managed to secure one million Norwegian kroners from the Norwegian Cultural Heritage Foundation for a number of projects.

9. See http://www.biotope.no.

10. See http://www.biotope.no/2012/10/birding-destination-varanger-pro-nature.html.

11. FFB includes Joar Nango, Eyvind Tallerås, and Håvard Arnhoff.

12. Also known as Persian hogweed.

13. Lucy Lippard, *Undermining: A Wild Ride through Land Use, Politics, and Art in the Changing West* (New York: The New Press, 2013), 94.

14. Lippard, "Peripheral Vision," 338.

15. Ibid., 339.

16. Ibid., 338.

17. Frans Berkhout, "Anthropocene Futures," *The Anthropocene Review* 1, no. 2 (2014): 157.

18. Fox, "The Art of the Anthropocene."

19. Peter Arbo, Audun Iversen, Maalke Knol, Toril Ringholm, and Gunnar Sander, "Arctic Futures: Conceptualizations and Images of a Changing Arctic," *Polar Geography* 36, no. 3 (2013): 163–182.

20. Henri Lefebvre, *La production de l'espace* (Paris: Éditions Anthropos, 1974).

21. Miwon Kwon, *One Place After Another: Site-specific Art and Locational Identity* (Cambridge, MA: MIT Press, 2004), 159.

22. Ibid., 159.

23. The authors are researchers in the Future North project (www.futurenorth.no).

Bibliography

Aarsæther, Nils, Larissa Riabova, and Jørgen Ole Bærenholdt. "Community Viability." In *Arctic Human Development Report*, edited by Nils Einarsson, Joan Nymand

Larsen, Annika Nilsson and Oran R. Young. Akureyri: Stefansson Arctic Institute, 2004.

Arbo, Peter, Audun Iversen, Maalke Knol, Toril Ringholm, and Gunnar Sander. "Arctic Futures: Conceptualizations and Images of a Changing Arctic." *Polar Geography* 36, no. 3 (2013): 163–182.

Berkhout, Frans. "Anthropocene Futures." *The Anthropocene Review* 1, no. 2 (2014): 154–159.

Ellsworth, Elizabeth, and Jamie Kruse, ed. *Making the Geologic Now: Responses to Material Conditions of Contemporary Life*. New York: Punctum Books, 2013.

Fox, William L. "The Art of the Anthropocene." In *Curating the Future: Museums, Communities and Climate Change*, edited by Jennifer Newell, Libby Robin, and Kirsten Wehner. Honolulu: University of Hawai'i Press, 2016.

Kwon, Miwon. *One Place After Another: Site-specific Art and Locational Identity*. Cambridge, MA: MIT Press, 2004.

Lefebvre, Henri. *La production de l'espace*. Paris: Éditions Anthropos, 1974.

Lippard, Lucy, "Peripheral Vision." In *Land Arts of the American West*, edited by Chris Taylor and Bill Gilbert, 337–345. Austin: University of Texas Press, 2009.

———. *Undermining: A Wild Ride through Land Use, Politics, and Art in the Changing West*. New York: The New Press, 2013.

McHarg, Ian. *Design With Nature*. New York: Doubleday, 1971.

Robin, Libby, Dag Avango, Luke Keogh, Nina Möllers, Bernd Scherer, and Helmuth Trischler. "Three Galleries of the Anthropocene." *The Anthropocene Review* (September 2014). doi: 10.1177/2053019614550533.

"The Shtokman office in Teriberka." *Barents Observer*, June 9, 2009, http://barentsobserver.com/en/node/18632.

9

Knowing Land, Quantifying Nature

Assessing Environmental Impacts in the Sahtu Region, Northwest Territories

carly dokis / NIPISSING UNIVERSITY

THE CANADIAN NORTH IS OFTEN PORTRAYED AS AN AREA OF PRISTINE wilderness and as a region of untapped natural resources. It is likewise characterized as a place contested by multinational corporations seeking to extract resources from the land, and indigenous peoples struggling to preserve and protect homelands and a subsistence way of life. Yet this dichotomous presentation of the Canadian North as either a pawn in a global capitalist agenda or a place where indigenous peoples are steadfast in preserving tradition mischaracterizes the ways in which decisions are locally made about who ought to use the land and for what purposes. In this chapter I explore how Dene people in the Central Mackenzie Valley consider the risks and impacts of extractive industries on their territories, and how these frequently conflict with rationalized forms of institutionalized environmental assessment. Drawing on nine months of fieldwork carried out in three Sahtu communities between 2006 and 2010, and a review of environmental assessment documents associated with the Mackenzie Gas Project,[1] I show sharp contrasts between how Sahtu Dene people know and experience their land and the ways in which nature is represented in institutionalized environmental assessment processes.

Sahtu Dene people have a deep history of adjusting to environmental and economic change, and in many ways balancing subsistence and extractive economies is not seen as contradictory or unexpected. Yet these diverse and changing land uses have not resulted in substantial alterations in the valuation of land as a means of sustenance that is material and emotive, physical and spiritual. Local decisions about land use are concurrently influenced by complex and unpredictable sources, including economies, translocal networks, and technologies, and by very localized sources and expectations such as prophecies and the desire to be a good human being. This commingling of what are often thought of as distinct spheres of influence—those of so-called development and tradition, or of capitalism and subsistence—includes more than just an intersection of "mixed economies" but carries with it entanglements in epistemological orientations, human and other-than-human relationships, and the subjective experience of nature. I argue here that it is precisely the emotive and life-sustaining experience of land for Sahtu Dene people that is rendered negligible in the institutionalization and subsequent quantification of the impacts and risks of extractive industries. That is, through the construction, technocratization, and subsequent management of industrial impacts as practiced in current institutionalized assessment processes, the social, cultural, and political risks of northern resource development for Sahtu Dene people are hidden from view. As a consequence of this concealment, the disproportionate distribution of risks associated with intensified extractive industries is not only minimized but becomes mischaracterized as one that can be remedied with economic solutions.

As I show here, the ways in which Sahtu Dene people engage in land-based decision-making processes do not preclude the use of the land for extractive projects; in fact, Sahtu Dene have at times been proponents of such endeavors. Yet, because decisions about land are always produced in arenas of power, the distinctly local means of making decisions, and the reasons behind them, often go unacknowledged by formal and rational regulatory institutions designed to predict and manage the impacts and risks of extractive industries. This seems to suggest that conflicts surrounding the nature and scope of extractive industries on Sahtu lands may have less to do with contestations between subsistence and extractive economies and rest more fundamentally in questions about sovereignty and the ability to make decisions that are reflective of community visions and experiences.

"Sahtu Is the Boss"

The Sahtu Region, as it exists today, was created as a result of the Sahtu Dene and Métis Comprehensive Land Claim Agreement, executed between the Dene and Métis in the Central Mackenzie Valley, the government of the Northwest Territories, and the government of Canada in September 1993. This comprehensive claim exchanged legally undefined Aboriginal rights in the region for legally recognized rights in 283,171 square kilometers of the newly created Sahtu Settlement Area (SSA). These rights include 41,437 square kilometers of land in fee simple title, rights to hunt and fish, the exclusive right to trap, federal government capital payments over a fifteen-year period, and participation in regulatory and co-management systems designed to involve Sahtu Dene and Métis people more directly in land use and environmental decision-making. Institutionalized Dene and Métis involvement in land use decision-making comes in a variety of forms, ranging from formal consultation between proponents, government, and community and regional governance bodies and the appointment of Sahtu Dene and Métis representatives on co-management boards, to more widespread participation from community membership in community meetings and environmental impact assessment hearings. As I have indicated elsewhere, the creation of this complex network of institutions and bodies designed to consider land use decisions and their associated impacts has had significant consequences for Sahtu social, political, and cultural relations, not the least of which includes the commodification of land under a corporate structure of land management and a rearranging of normative land use decision-making processes and protocol.[2]

Sahtu Dene people have been living in the Central Mackenzie region for as long as anyone can remember; they say that they have been living here since the world was new.[3] Their stories and places upon the landscape recall the many generations of intimate association between the people and their land. They continue to tell these stories to their young people and to pass on the knowledge and skills of their Elders. Over the past half century, people in the Sahtu have witnessed a change from life lived primarily on the land to residence in permanent communities ranging in size from approximately one hundred to six hundred and fifty people. Many of the Sahtu Dene youth who have grown up in these communities have never spent an entire winter season in the bush, although there are Sahtu communities, such as Colville Lake, where entire families, children included, go out to their trap lines in late September and do not return to the village until Christmas. Regardless of whether or not people are engaged in land-based pursuits full or part time, Sahtu Dene people express a deep reverence for the land and a knowledge of the particular kinds of relationships that are required to sustain the unique place that Sahtu Dene people have in their world.

For Sahtu Dene, land, and consequently land-based practices, are seen as inherently changing and in flux. Among many people in the Sahtu, there is a belief in a distant time when significant events took place that now explain the characteristics of the modern universe. At this time, the world was inhabited by giant animals and people who could transform into other beings. It was often through the actions of these ancient beings that the world, as it now exists, was created.[4] These stories are etched in the landscape, recalled and re-experienced through the course of spending time and traveling on the land, and contain principles of proper behavior and action.[5] One such story involves Yamoria, one of two brothers who came into the world when the world was new and set everything in its place. On one occasion, Yamoria saved the people living near Great Bear Lake from flooding when three giant beavers built a dam on the Great Bear River. The giant beavers had built their dam near what today is known as the St. Charles Rapids, threatening to drown all of the human and other-than-human beings living upriver. At that time, Yamoria killed the large beavers that built the dam, cooked their meat on "the smokes"[6] (a burning coal seam) just southwest of Tulit'a, and hung the beaver pelts on Bear Rock. Yamoria shot two large arrows from Bear Rock into the water where the Mackenzie and Great Bear Rivers meet. These two large poles, called Yamoria's Arrows by local people, are visible at this location even today, as are the beaver pelts that hang as large colorful markings on the top of Bear Rock.

In this way, the environment, seen as being composed of animate and powerful beings, does not submit to human management, but rather it is the land and the agentive beings who dwell there that determine the fate of human beings. While I was living in Déline, I frequently had the opportunity to join local people on the land to travel and to hunt, fish, and gather berries and other land-based forms of sustenance. When asking if someone was going out across Great Bear Lake to hunt, fish, or gather, the most common reply was "*Sahtu k'aowe*," which roughly translates to "the Sahtu is the boss."[7] So, for Sahtu Dene, it is the land that determines the timing and pace of land-based activities not the intentions and planning of human beings. Rather than an objectified nature to be managed and controlled by humans through scientific rationality, for Sahtu Dene the land is seen as a leader (*k'aowe*) that is known through intimate sensory experience and knowledge that is handed down through generations by other beings who have also spent time there. And, for Dene, maintaining a proper relationship with the land is secured not through "managing" nature but by upholding Dene Law,[8] by spending time on the land and learning the old-timer stories, and by subsequently acquiring sensory-based knowledge.

I became acutely aware of the power of the land while traveling by boat from Tulit'a to Déline in early autumn of 2006. Tulit'a, a predominately Dene community of 450 people located at the intersection of the Mackenzie and Great Bear Rivers, had been hosting a four-day hand-game tournament, and people from all over the Sahtu had gathered there to participate in hand games, dancing, bingo, talent shows, and a general rekindling of social bonds. At the end of the tournament, I had secured a ride back to the community I was staying in with John,[9] a young man from Déline who was known for his exceptional bush skills in spite of his young age. For the price of gas, about $40, John had agreed to take me with him in one of several boats that were making the trip up the Great Bear River and across the Keith Arm of Great Bear Lake to the community of Déline. We left Tulit'a at about nine o'clock in the morning and began a rather smooth trip in good weather up the river. However, at the St. Charles Rapids we encountered some difficulties when one of the boats we were traveling with could not make it up the rapids because they were not deep enough for a speedboat to navigate safely. John took myself and a handful of other passengers up the river a short distance and dropped us off at the shore so that he could return and tow the speedboat through the rapids. I quickly realized how ill prepared I was for this trip. I had packed no food or sleeping bag and had no means of protecting myself from bears or wolves. I did not know which trees were good for firewood or where (or how) I could obtain them. In short, I had no relationship with this land, and thus had no knowledge of how to survive there. This was in sharp contrast to the Dene people with whom I was traveling. They quickly got wood and made a fire, and out of their packs they produced a kettle to make tea and a rack for cooking. Soon enough, someone pulled out some caribou ribs and a sharp knife. As was often the case in the Sahtu, people were generous with their time and their resources, and showed me a little about their profound knowledge of the land that allows them to make a comfortable living in one of the harshest environments on earth. As we sat along the shoreline and feasted on caribou and tea until the rest of the party traversed the rapids, one traveler said, "We know the importance of the land, and that it is all connected. People that come in and do the exploration, they don't know much about the land because they have not spent time here. They have not fished or hunted or trapped here. It is so different for us Dene. People here still use the land, and so what they know and say is really true."[10] In this way, proper Dene relationships with the land are tied not only to action but also to the sensory experience of place, which ultimately leads to the acquisition of knowledge.

The trip up the Great Bear River (which had taken a mere four hours to travel down on the way to Tulit'a) took eleven hours to reach the head of the river. By this time it was dark and cold, the wind was picking up, and we still had to cross the often-turbulent Keith Arm of Great Bear Lake. Three boats, including our own, sat on the shore of the lake where it met the river. Directly across the lake from us sat the community of Déline. There was a lot of debate between the drivers of the boats about whether or not we should attempt to cross the lake. It was potentially dangerous—the waves had picked up and we could not immediately tell how choppy it was in the middle of the lake. Most people began unpacking their boats and getting ready to spend a chilly subarctic autumn night out on the land waiting out the wind. Finally, about an hour later another boat came up the river with four or five experienced bushmen. They had a discussion in Slavey[11] about the conditions on the lake and determined that it was safe enough to cross. Once we decided to cross the lake, some of the women instructed me that I should go with them to the water's edge to "pay the water" by offering tobacco or money or another item of value. This offering, they said, would serve as a request that the lake allow us to cross it in safety. After arranging the boats so that there were at least two experienced drivers in each one, we took off across the lake. The other boats followed. We used lighters and flashlights to signal the other boats in the massive waves and were guided by a full red moon. At one point, we circled back toward the river as we lost sight of one of the boats and could no longer see their lights. After a very harrowing twenty minutes, we arrived safely in Déline. When we arrived we were greeted by a number of community members at the Little Lake, where our boats came ashore. They had seen our lights out on the lake and were waiting and ready to welcome us home. Our safe passage over the lake was dependent on the long-term experiential practice that involved sensing the lake (even when we couldn't see) that the drivers had acquired from years of experience reading and responding to the lake and from "paying the water" and thus maintaining appropriate relationships with that particular place.

At the heart of Sahtu Dene conceptions of morality and relationships with other-than-human beings is the concept of respecting the land. This goes beyond a simple consideration for not contaminating the landscape or a care not to overhunt or overfish to a profound reverence for the life-giving properties of the earth, water, animals, weather, insects, fire, fish, birds, and all other components of the other-than-human world. Dene people talk about the land as a provider of physical, spiritual, emotional, and intellectual sustenance. The land provides the physical necessities of life in the form of food and materials; it offers spiritual fulfillment because it is on the landscape where the acquisition and experience of power is most profound;[12] it sustains emotional connections to people, ancestors, places, and "being Dene"; and it is through primary experience in the bush that knowledge can come to be regarded as true. According to Sahtu Dene people, the land provides all these things at no cost. As one Dene hunter said to me, "We and the animals do not have to pay the land. The birds, and moose, and caribou and all of the other animals have places to live on the land and they follow a universal law and do not break it. We can live on the land and it is free, just so long as we follow the law."[13] Beliefs about respecting the land and upholding Dene law are about the protection of land for subsistence livelihoods, but they are also deep commitments to protecting the very fabric of what it means to be a Dene person.

While Sahtu Dene people remain fundamentally engaged with a landscape that they see as agentive and implicated in reciprocal and kinlike relationships with humans, they also incorporate new technologies and approaches to land use that reflect changing economies and subsistence practices. For example, when I was conducting my fieldwork in 2006, there was a move by three Dene-owned businesses to construct a hydroelectric dam on the Great Bear River that would then sell power to the operators of the Mackenzie Gas Project. This proposed dam would be located at the St. Charles Rapids, the very same place where the giant beavers made their dam when the world was new. The proposed dam never moved past the planning stages, not because it was met with resistance from local people but because the operator of the Mackenzie Gas Project, Imperial Oil, was uninterested in purchasing hydroelectricity. It is important to note that it was the Dene themselves who were proponents of the dam, and that resistance to the dam by local people, in spite of its location on a place of ancient significance, was indeed very minimal. In some cases, when I asked local residents about their position on the potential for the construction of a dam on the Great Bear River, they responded that they did not need to worry about a dam on the river because there was no prophecy about the dam, and therefore the dam would never materialize.[14]

Likewise, the community of Colville Lake, where the Dene refused water treatment technology and wastewater infrastructure until 2012 because they believed that this technology was inconsistent with Dene values of independence and spending time on the land, was one of the only communities in the Sahtu that had high-speed Internet access during my fieldwork; they used it to maintain communications with oil and gas companies in Calgary who were looking to develop gas deposits on lands where the Dene hold subsurface leases. Thus, the use of new technology and changing land use patterns are not haphazard or careless but reflect Dene conceptions of the land (and consequently land-based practices) as constantly changing and fluctuating. Adaptation to the changing nature of the landscape is seen as an intrinsic component of life and has been required over time to deal with changing ecological and economic circumstances, and will continue well into the future.[15] These adaptations, however, also reflect consistencies in how Sahtu Dene know and value their land. For example, when anthropologist Scott Rushforth conducted his fieldwork in and around Déline in the 1970s, during a time when the first Mackenzie Valley Pipeline was proposed,[16] he noted that many local people were not necessarily opposed to extractive industries on Sahtu lands so long as the domestic economy was maintained and protected.[17] The same was true when I conducted my fieldwork some thirty years later. As one community leader said, "The Elders use the land in one way, and the younger generations use it in the same, but in different ways. Young people usually go out on the weekend and use snowmobiles instead of dog teams. Fish lakes were really important when people used dog teams, but younger generations do not use them that way anymore. Both Elders and young people have really important knowledge about the land that is valuable. But for younger generations people need jobs. As long as we can control what development happens, we can provide jobs and still go out on the land."[18]

This suggests that for Sahtu Dene the preservation of a land-based subsistence economy and the development of extractive economies are not inherently mutually exclusive, particularly if the intensity and extent of extractive economies are controlled by the Dene themselves. Even for Sahtu Elders, who are proficient and steadfast teachers about the importance of maintaining respectful and reciprocal relationships with their land in order to preserve a Dene way of life, there is an acknowledgment that the young people need work and must simultaneously engage in wage labor in order to support themselves in a changing world. As such, debates that occasionally arise between divergent forms of land use (often between intensive resource development and the preservation of a domestic economy) stem less from changing forms of land practice and, as I will argue below, rest more fundamentally on how the impacts of such activities are differently known and evaluated.

Predicting and Creating Futures: The Management of Industrial Impacts in a Risk Society

For Dene the land is known through sensory perception and valued not only for the sustenance that it provides but for the emotive, subjective, and embodied experience of place. In contrast, for environmental managers seeking to predict the impacts of extractive projects, knowledge about the land is sought in processes of scientific rationality, which emphasize calculability, efficiency, predictability, and control of uncertainty through the production of technocratic quantitative indicators and measurements. Historian and geographer Dean Bavington has shown that managerial ecology has its roots in sixteenth- and seventeenth-century Europe, where scientific, economic, and political processes resulted in a transformation of sensory knowledge (and valuation) of nature into the control of nature through increasingly bureaucratic forms that have close ties with capitalist production.[19] In this way, the expansion of managerial ecology has coincided with other forms of colonial expansion and frames the ways in which the impacts and risks of extractive processes are assessed and evaluated. Importantly, as Bavington notes, the exercise of managerial powers is not restricted to statecraft but is also essential for corporations "interested in conquering, instrumentalizing, and commodifying space."[20]

An essential component of managerial ecology involves the control and maximization of productive conditions through the prediction and management of ecological risks.[21] In the Sahtu, the influence of managerial ecology in land-use decision-making is significant, in spite of attempts to integrate what is often politically classified as "traditional knowledge" into environmental regulatory and decision-making boards.[22] This knowledge—often translated into quantitative data that reveals the number and kinds of animals, fish, and birds harvested by Sahtu Dene and Métis hunters, fishers, and trappers—is collected on an ongoing basis and is used by co-management boards, governments, and industries in land-use decision-making—both for protecting harvesting traditions and for evaluating and monitoring the impacts of extractive projects. Yet the management of ecological conditions is simultaneously the management of other realms of social life: of people, ideas, and relationships. So, in this way, the objectification of nature through managerial processes immediately frames environmental (and social) issues as ones that can be addressed through rational control over nature. Under managerial forms, relationships that Sahtu Dene people have with their land are subverted precisely because they resist rational quantification.

By way of illustration, consider the way in which the Environmental Impact Statement (EIS) for the Mackenzie Gas Project presents an objectified and quantified nature in its assessment of the impacts of a natural gas pipeline. First, the environment is dissected into separate parts that can be measured and counted—these parts are known as '"valued ecosystem components." The means of identifying the "valued wildlife components" are presented in fig. 9.1,[23] and the assessment of the risks associated with each "valued wildlife component" is demonstrated in fig. 9.2.[24] The rational assessment of ecosystem components portrays a manageable nature, a nature that is devoid of animacy and of agentive properties and is rendered knowable through technocratic measurements and models. Constructed in this way, emotive experiences of nature—as a place that gathers stories and personal experience and is embedded with willful and subjective properties and the associated means of knowing land through sensory experience—are rendered invisible through the violence of quantitative statistics and numerical values.

Anthropologist Clinton Westman has been critical of the ways in which social impact assessments consider and assess the impacts and risks of extractive industries on indigenous cultures. Citing a lack of empirical and methodological rigor consistent with sound anthropological practice, Westman argues that there is a systematic avoidance of topics that are not easily rendered technical, and this serves to dismiss "spiritual, cosmological, or ontological issues at the heart of indigenous cultural traditions."[25] Anthropologists not only work to connect seemingly unrelated realms of life to a more (or less) integrated sense of the world but also seek to understand connections between subjects and contexts, usually through long-term participant observation. This process invariably leads to understanding culture as rooted in systems of meaning that shape how people see, think about, and engage with the world. Take, for example, Clifford Geertz's notable description of culture as "webs of significance" and his insistence that "getting at" culture is "not an experimental science in search

Section 10: Wildlife
Significance of Potential Effects of the Pipeline Corridor on Wildlife Habitat Availability

Valued Component	Phase When Impact Occurs	Effect Attribute				Significant
		Direction	Maximum Magnitude	Maximum Geographic Extent	Maximum Duration	
Barren-ground caribou	Construction	Adverse	Low	Local	Far future	No
	Operations	Adverse	Low	Local	Far future	No
	Decommissioning and abandonment	Adverse	Low	Local	Far future	No
Woodland caribou	Construction	Adverse	Low	Local	Far future	No
	Operations	Adverse	Low	Local	Far future	No
	Decommissioning and abandonment	Adverse	Low	Local	Far future	No
Moose	Construction	Adverse	Moderate	Local	Medium term	No
	Operations	Adverse	Low	Local	Far future	No
	Decommissioning and abandonment	Adverse	Low	Local	Far future	No
Grizzly bear	Construction	Adverse	Moderate	Local	Long term	No
	Operations	Adverse	Low	Local	Long term	No
	Decommissioning and abandonment	Adverse	Low	Local	Long term	No
Marten	Construction	Adverse	Low	Local	Far future	No
	Operations	Adverse	Low	Local	Far future	No
	Decommissioning and abandonment	Adverse	Low	Local	Far future	No
Lynx	Construction	Adverse	Low	Local	Long term	No
	Operations	Adverse	Low	Local	Long term	No
	Decommissioning and abandonment	Adverse	Low	Local	Long term	No
Beaver	Construction	Adverse	Low	Local	Long term	No
	Operations	Adverse	Low	Local	Long term	No
	Decommissioning and abandonment	Adverse	Low	Local	Long term	No
Amphibians	Construction	Adverse	Low	Local	Medium term	No
	Operations	Adverse	Low	Local	Medium term	No
	Decommissioning and abandonment	Adverse	Low	Local	Medium term	No
Birds	Construction	Adverse	Low	Local	Far future	No
	Operations	Adverse	Low	Local	Far future	No
	Decommissioning and abandonment	Adverse	Low	Local	Far future	No

Fig. 9.1. Assessment of Environmental Impacts on Ecosystem Components. SOURCE: MACKENZIE GAS PROJECT (2004, 5:SECTION 10, 214)

Table 10-5: Wildlife Value Components–Pipeline Corridor

Valued Component		Regulatory Status	Ecological Importance	Socio-Economic Importance	Conservation Concern
Latin Name	Common Name				
Rangifer tarandus groenlandicus	Barren-ground caribou	-	•	•	•
Rangifer tarandus caribou	Woodland caribou	•	•	•	•
Alces alces	Moose	-	•	•	•
Ursus arctos	Grizzly bear	•	•	•	•
Martens americanus	Marten	-	•	•	-
Lynx canadensis	Lynx	-	•	•	-
Castor canadensis	Beaver	-	•	•	-
n/a	Amphibian community	•	•	-	-
Chen caerulescens	Snow goose	-	•	•	•
Aythya marila and Aythya affinis	Scaup (greater and lesser)	•	•	•	•
Falco peregrinus	Peregrine falcon	•	•	-	•
Tringa flavipes	Lesser yellowlegs	•	•	-	-
Stema paradisaea	Arctic tern	-	•	-	-
Poecile hudsonicus	Boreal chickadee	•	•	-	-

NOTES:
n/a = not applicable
1 The socio-economic factors in selecting VCs included subsistence harvest, other hunting and other recreational and guiding and outfitting interests
• Indicates item was a factor in selecting VC
- Indicates item was not a factor in selecting VC

Fig. 9.2. Valuation of Ecosystem Components. SOURCE: MACKENZIE GAS PROJECT (2004, 5:SECTION 10, 17)

of law but an interpretive one in search of *meaning*."[26] For anthropologist Eric Schwimmer, one of the main orientations of anthropology is an exploration of the ways in which subjects make sense of local contexts and, consequently, as John Clammer notes, "the ways that local subjects find themselves through the notion of ontology, understood as the systems of signs and objects used to construct local theories of being in the world."[27] This (re)focusing on connections between contexts and their subjective experience by local actors provides a way of exploring how people make sense of contradictions and inconsistencies such as modernity and tradition,[28] and can offer an analysis of the ways in which local negotiations of global forces can serve to resist totalizing tendencies of marketization and commodification.

In contrast to anthropological methodologies that stress the interpretation of cultural meanings, rationalized environmental assessment practices develop ways in which culture can be registered and subsequently valued. In attempting to predict the impacts of pipeline activities on "traditional culture" in the Sahtu, the EIS for the Mackenzie Gas Project operationalizes culture as an *action* in order to establish indicators that can be quantified and measured:

> Indicators of culture can be seen in people's behavior. Culture is reflected in activities that are shaped by beliefs and values, activities that make use of traditional knowledge, skills and disciplines. What people do and are able to do thus serve as indicators of their involvement in traditional cultures. The following activity-based indicators are used (to measure involvement in culture) . . . involvement in traditional harvesting, the amount of country food consumed, the ability to speak a traditional language.[29]

Table 5-16: Adults Who Hunted or Fished in the Sahtu Communities

Location	1993 (%)	1998 (%)	2002 (%)
Northwest Territories	18	42	41
NWT Aboriginal communities[1]	42	48	51
SSA total	28	45	48
Norman Wells	8	44	38
Sahtu Aboriginal communities total	37	46	53
Fort Good Hope	33	39	-
Déline	41	53	-
Tulita	32	45	-
Colville Lake	71	56	-

NOTES:
- = data not available
1 All study area communities in the Northwest Territories except Inuvik, Norman Wells, Fort Sampson, Yellowknife, Hay River and Enterprise
Percentage of population aged 15 years and older

SOURCE: GNWT Burea of Statistics (1999, 2002b)

Fig. 9.3. Indicators of involvement in "traditional culture" according to land use. SOURCE: MACKENZIE GAS PROJECT (2004, 4:SECTION 5,16)

Using these indicators to determine "baseline conditions," the environmental impact statement goes on to measure involvement in traditional culture in numerical terms (see figs. 9.3 and 9.4).

One of the reasons for identifying impacts in such a manner is to make them predictable, calculable, and, once identified, manageable. When risks or impacts associated with extractive industries are identified through rationalized assessment processes, the control of these risks is often sought through technological solutions. For example, mitigation strategies proposed in the EIS for the remedy of impacts on wildlife range from developing operating guidelines that manage access to wildlife resources at work sites to implementing reclamation plans to reestablish wildlife habitat.[30] These proposed remedies for potential risks and impacts of extractive industries are not reflective of metaphysical views that Sahtu Dene people have of the land and the beings who dwell there as agentive and implicated in social ties that connect humans and other-than-human beings together and to the land. As a community member from Déline stated before the Joint Review Panel for the Mackenzie Gas Project in a community hearing, "And another thing you mentioned about the study of animals, monitoring. Fish, animals, ducks, everything,

Table 5-22: Aboriginal Language Speakers in the Sahtu Communities

Location	1989 (%)	1994 (%)	1999 (%)
Northwest Territories	56	50	45
SSA total	86	68	64
Norman Wells	52	36	29
Sahtu Aboriginal communities total	88	73	68
Fort Good Hope	81	54	48
Déline	98	96	93
Tulita	82	61	63
Colville Lake	95	96	76
NOTE: Percentage of Aboriginal people, aged 15 years and older			
SOURCE: GNWT Burea of Statistics (2003e)			

Fig. 9.4. Indicators of involvement in "traditional culture" according to language use. SOURCE: MACKENZIE GAS PROJECT (2004, 4:SECTION 5,22)

Table 5-19: Replacement Values of Wildlife Harvests by Sahtu Harvesters

Species	1999		2001	
	Total Harvest (kg)	Harvest Value ($)	Total Harvest (kg)	Harvest Value ($)
Mammals	171,165	2,375,936	134,462	1,866,471
Birds	8,286	146,753	6,577	116,482
Fish	42,613	686,086	20,977	337,733
Total	222,064	3,208,775	152,016	2,320,686
NOTE: See the text for details about deriving replacement costs per kilogram for mammals, birds and fish				
SOURCE: Sahtu Renewable Resource Board (1999, 2001)				

Fig. 9.5. Valuation of hunting, trapping, and fishing. SOURCE: MACKENZIE GAS PROJECT (2004, 4:SECTION 5,19)

different kinds of animals, if there's an oil spill, how are they going to know it? But they're still going to drink the water if there's still oil on the land; then too, they're going to suffer and die from it, too."[31] As is evident here, the application of technical procedures associated with remedying ecological hazards does not fundamentally take into account the *systems of meaning* that are entangled with Sahtu Dene land-based practices and relationships to land, nor do they restore these relationships once they have been severed.[32]

When rationalized assessment processes acknowledge that an impact of extractive projects cannot be managed through technological means, compensation for such an effect is rendered in economic terms. So when the EIS for the Mackenzie Gas Project identified that there is a risk of adverse impacts on hunting and fishing that could not be managed through technological means, the EIS calculated the monetary replacement value of each animal: "a monetary value for the SSA wildlife harvest was calculated based on appropriate weighting of the cost of replacing the edible mammal, bird, and fish with store-bought beef, chicken, and fish." Fig. 9.5 shows the calculation of the value of hunting and fishing.[33]

Evaluated in this way, the EIS translates the subjective and embodied experience of hunting and fishing, and their wider subjective meanings, into quantified values that can be replaced through economic compensation. It technocratizes and commodifies entire realms of life that are differently valued, experienced, and known by Sahtu Dene people, and reduces an agentive and affinal landscape, and systems of meanings used to engage it, to an exchange value in a market that can make Sahtu Dene experiences of hunting commensurable with a trip to the grocery store. This quantification simultaneously serves to subvert localized systems of meaning that are not singly economic in nature but also characterize how people think about the world and their place in it.

Constructing a Nature Without People

In *Risk Society: Towards a New Modernity*, Ulrich Beck argues that contemporary debates surrounding the evaluation of environmental risk are predominately framed by the application of natural sciences, which is legitimated in part because of their ability to identify hazards that escape sensory perception.[34] Yet the problem of framing debates about ecological or social risks in technocratic terms that objectify and quantify human-nature relations is that they have the potential to render the subjective experience of land as a secondary form of understanding the world. As Beck writes, "there exists accordingly a danger that an environmental discussion conducted exclusively in chemical, biological and technological terms will inadvertently include human beings in the picture only as *organic material* . . . it runs the risk of atrophying into a discussion of nature *without* people, without asking about matters of social and cultural significance."[35] In this way, the risks and impacts associated with industrial or extractive projects become mapped as indicators or quantifiable measurements on the landscape, without a proper accounting of how these risks and impacts are differently experienced by the people whom they effect. As Beck notes, the rationalized assessment of risk assumes "that all people everywhere are *equally* affected"[36] by ecological or industrial hazards, regardless of whether or not people hold different environmental, social, and cultural experiences of land and land-based activities.

In the Sahtu, the marginalization of the subjective experience of nature in assessing the risks and impacts of extractive industries is simultaneously a marginalization of components of life that people associate with being Dene. As one community leader said in a hearing on the pipeline:

> We are very grateful that we call ourselves Sahtu Dene, Bear Lake people, and I'm very proud to be a leader for the Great Bear Lake people, but I'm very concerned that the protection of my culture is not addressed seriously in these hearings or through the different regulatory departments or bodies that were set up to help us . . . I want to make sure that we can have a balance when it comes to development and protecting our way of life in the North here . . . I want to make sure that these kids aren't forced to go down one path, which is a wage economy.[37]

As shown here, the entanglements between realms of life that have been constructed as oppositional—between so-called tradition and modernity, or between subsistence and market—seem to have less to do with an emphasis on preserving a static past but rest more fundamentally in Sahtu Dene concerns about their authority to determine the extent and nature of the risks and impacts of extractive industries and to influence land-use decision-making in ways that take into account the cultural meanings of associated land-based practices. As currently undertaken, attempts to predict and manage the risks and impacts of extractive projects in rationalized assessment processes serve to measure and quantify what is for Sahtu Dene unquantifiable—those subjective meanings that people have of their relationships to land—and to make them exchangeable for economic compensation. To be sure, Sahtu Dene people have drawn, and will continue to draw, material substance from their lands, but Sahtu Dene land-use decision-making also draws on noneconomic influences such as prophecies, embodied experiences of place, and what it means to be a good human being. By framing the risks and impacts of extractive industries in economic terms only, localized experiences of the spiritual, emotive, and ontological importance of the land and land-based practices then become implicated in commodity markets that cannot properly account for their subjective values.

In large parts of the Canadian North, public involvement in the regulation and assessment of resource extraction and other environmental problems has become a common expectation of persons and communities whom it might affect. This is particularly the case for indigenous communities in Canada who have a legal right to be consulted about activities that may infringe upon their rights and impact their lands.[38] Yet there has been little evidence to show that this increased involvement has translated into meaningful dialogue, where the statements made by indigenous persons have been heard on their own terms. What is more, there is growing criticism, both from within Canada and internationally, that what these forms of "participation" or "consultation" actually produce are renewed configurations of coloniality whereby local "theories of being in the world," and connections between subjects and contexts, become digested by bureaucratic regulatory apparatuses and spewed out as something disarticulated and wholly unfamiliar. In fact, recent frustration by indigenous peoples in Canada, largely as a result of intensified resource extraction coupled by environmental deregulation, has resulted in the Idle No More movement, one of the largest indigenous movements in Canada since the 1970s.[39]

While proponents and regulators alike have focused on achieving scientific certainty about the prediction of environmental impacts, there has been comparatively less regulatory dialogue on the uncertainty inherent in mobile capital and international commodity markets and the "risks" that this uncertainty poses to the viability of extractive projects and to socioeconomic conditions in local communities.[40] It is somewhat ironic then that, in the case of the Mackenzie Gas Project, it was world commodity prices and a record low price of natural gas that ultimately derailed the building of the pipeline. This was, of course, after communities in the Sahtu had been promised jobs and were gearing up—establishing training programs, starting small businesses—for an intensified hydrocarbon-based economy. The plans for new jobs, new daycares, and new forms of cash-based revenue were put to the side, or abandoned altogether, as the dream of building a pipeline through the Mackenzie Valley faded. Given their experiences with changing economies, perhaps this wasn't much of a surprise for Sahtu Dene people. Yet it does speak to the very important reasons why there is little trust by Sahtu Dene people that an economy based on hydrocarbon resources alone can be managed by technological or economic means. As I was reminded by a Sahtu Dene elder, "Without money we can live on the land, but without the land we can not live on money."[41]

Endnotes

1. The Mackenzie Gas Project (MGP) is a 1,220-kilometer natural gas pipeline and related infrastructure proposed for the Mackenzie Valley by Imperial Oil, ConocoPhillips Canada, Shell Canada, ExxonMobil Canada, and the Aboriginal Pipeline Group. The MGP would develop three onshore gas fields in the Mackenzie Delta and transport natural gas and natural gas liquids from processing facilities near Inuvik to existing Alberta pipelines. The proponents submitted regulatory applications to construct and operate the $16.2 billion pipeline in 2004, including an Environmental Impact Statement (EIS). The MGP went through one of the largest environmental assessments in Canadian history, with the Joint Review Panel (JRP) for the Mackenzie Gas Project mandated to consider the environmental and socio-economic impacts of the pipeline and the National Energy Board (NEB) responsible for regulatory hearings (tolls, tariffs, engineering, and design). In 2010 the regulatory authorities recommended that the pipeline be approved, and after the approval of the federal government, the NEB issued a Certificate of Public Convenience and Necessity to the pipeline proponents in 2011.

2. For a more detailed discussion of the effects of the comprehensive land claim on land-use decision-making in the Sahtu, see Carly Dokis, "Modern Day Treaties: 'Development', Politics, and the Corporatization of Land in the Sahtu Dene and Métis Comprehensive Land Claim Agreement," *Geography Research Forum* 30 (2010): 32–49.

3. For an account of Sahtu Dene old-timer stories, see George Blondin, *When the World Was New: Stories of the Sahtu Dene* (Yellowknife: Outcrop, The Northern Publishers, 1990).

4. Blondin, *When The World Was New*. See also George Blondin, *Yamoria: The Law Maker* (Edmonton: NeWest Publishers, 1997).

5. Scott Rushforth, "Legitimation of Beliefs in a Hunter-Gatherer Society: More About Bearlake Athapaskan Knowledge and Authority," *American Ethnologist* 21 (1992): 486.

6. What Sahtu Dene people call "the smokes" is a large burning coal seam located on the west side of the Mackenzie River, about four kilometers south of Tulit'a. The coal seam burns continuously, and according to locals, if a person is able to see the smoke rising from the coal seam, it means that the person will live a long life. It is said that this is the campfire made by Yamoria when he cooked the meat of the large beavers who built a dam on Great Bear River.

7. Sahtu Dene Elder Alfred Taniton makes a similar point in a story he shares in Raymond Taniton and Mindy Willet, *At the Heart of It: Dene dzó t'árá* (Markham, ON: Fifth House/Fizhenry & Whiteside, 2011).

8. Dene law, or *Dene ʔeʔah*, is a set of moral codes that one ought to live by in order to be a good human being. The main components of Dene law include sharing, mutual assistance, and maintaining respectful and mutually reciprocal relationships with the environment—for example, offering tobacco or payment for safe passage on the land or water or not wasting parts of animals that have offered themselves in the hunt.

9. A pseudonym has been used here to protect the confidentiality of the individual.

10. Fieldnotes, September 5, 2006.

11. Slavey is an Athabascan language spoken by Dene people across the western subarctic. A variety of Slavey dialects are spoken throughout Denendeh, or the Dene homeland, including *Sahtúot'įnę Yatį́*,or North Slavey, spoken in the Sahtu region.

12. Ellen Basso, "The Enemy of Every Tribe: Bushman Images in Northern Athabaskan Narratives," *American Ethnologist* 5, no. 4 (1978): 690–709.

13, Fieldnotes, August 25, 2006.

14. There is a strong prophecy tradition in the Sahtu whereby people with strong powers receive visions that enable them to help others and to foretell future events. Prophecy traditions have been well documented by anthropologists and others working among northern Athabascan peoples, including Jean Guy Goulet, *Ways of Knowing: Experience, Knowledge, and Power Among the Dene Tha* (Lincoln: University of Nebraska Press, 1998); Antonia Mills, "The Meaningful Universe: Intersecting Forces in Beaver Indian Cosmology and Causality," *Culture* 6, no. 2 (1986): 81–91; Robin Ridington, *Trail to Heaven: Knowledge and Narrative in a Northern Native Community* (Iowa City: University of Iowa Press, 1988).

15. Over the past two hundred years, Sahtu Dene have experienced significant changes on their lands and in their lives. Initial contact between the Dene and Europeans came as a result of the fur trade, with trading posts established in the Sahtu beginning in 1789. Anthropologist Michael Asch has argued that as late as the 1890s transportation challenges limited the quantity of goods that could be brought in and out of the Northwest Territories, and as a consequence, a great deal of trade actually consisted of Dene people providing food to traders in exchange for Western goods. The fur trade certainly had numerous consequences for the Sahtu Dene; however, historian Kerry Abel argues that though Dene understandings of acceptable trade relationships differed from those of the European traders, Dene participation in the fur trade was not only practical and economical but consistent with Dene lifeways and morality. For a detailed account of Dene participation in the fur trade, see Kerry Abel, *Drum Songs: Glimpses of Dene History* (Montreal: McGill-Queen's University Press, 1993), and for overall changes in Dene economies see Michael Asch, "The Dene Economy," in

Dene Nation: The Colony Within, ed. Mel Watkins (Toronto: University of Toronto Press, 1977).

16. In 1974, in response to interest by petroleum companies who had made significant natural gas discoveries in the Beaufort Sea and the North Slope of Alaska, the Department of Indian Affairs and Northern Development established a Commission of Inquiry, led by Justice Thomas Berger, into the feasibility of a gas pipeline through the Mackenzie Valley. In 1977, after three years of hearings, Justice Berger issued a report titled *Northern Frontier, Northern Homeland,* in which he recommended that the development of a pipeline through the Mackenzie Valley should be postponed for ten years in order to allow sufficient time for land claims to be settled and for new institutions and programs to be established.

17. Scott Rushforth, "Country Food," in Watkins, ed., *Dene Nation*.

18. Fieldnotes, September 20, 2006.

19. Dean Bavington, *Managed Annihilation: An Unnatural History of the Newfoundland Cod Collapse* (Vancouver: University of British Columbia Press, 2010).

20. Dean Bavington, "Of Fish and People: Managerial Ecology in Newfoundland and Labrador Cod Fisheries" (PhD diss., Wilfred Laurier University, 2005), 5.

21. Carolyn Merchant, *The Death of Nature: Women, Ecology, and the Scientific Revolution* (New York: Harper & Row, 1980).

22. For more on the complexities of institutionalized resource co-management in the Sahtu, see Carly Dokis, *Where the Rivers Meet: Pipelines, Politics, and Participatory Management in the Sahtu Region Northwest Territories* (Vancouver: University of British Columbia Press, 2015). For an excellent examination of the complications associated with the inclusion of TEK in environmental management, see Paul Nadasdy, "The Anti-Politics of TEK: The Institutionalization of Co-Management Discourse and Practice," *Anthropologica* 47, no. 2 (2005): 215–232.

23. Mackenzie Gas Project. *Environmental Impact Statement*, vol. 5, section 10, 10–17.

24. Ibid., 10–214.

25. Clinton Westman, "Social Impact Assessment and the Anthropology of the Future in Canada's Tar Sands," *Human Organization* 72, no. 2 (2013): 111.

26. Clifford Geertz, *The Interpretation of Cultures* (New York: Basic Books, 1973), 5.

27. John Clammer, "Decolonizing the Mind: Schwimmer, Habermas and the Anthropology of Post-Colonialism," *Anthropologica* 50, no. 1 (2008): 159.

28. Natacha Gagne, "The Link Between an Anthropologist and His Subject: Eric Schwimmer and (De)colonization Processes," *Anthropologica* 50, no. 1 (2008): 13.

29. MGP, vol. 4, sec. 5, 16.

30. MGP, vol. 4, sec. 5, 22.

31. Northern Gas Project Secretariat, *Joint Review Panel for the Mackenzie Gas Project—Public Hearing Transcripts*, vol. 16 (2006): 1636.

32. Clinton Westman, in a 2013 article examining social impact assessments associated with the Alberta tar sands, makes similar observations about the dismissal of the subjective experience of land in the assessment of extractive industries there. He notes that the knowledge generation processes associated with impact assessment serve to transform political questions about what ought to be done with the land into technical interventions that privilege scientific and rational management over other embodied ways of knowing and being. As is the case in the Sahtu, Westman shows that in the context of the Alberta tar sands, the evaluation of impacts associated with tar sands extraction often conclude that "while there may be some impact [to land-based activities] during construction and operations phases, in the long run, access to the activity will be *improved* following tar sands development and reclamation." He goes on to argue that the function and subsequent consequences of such assessments are to create and legitimate certain scenarios of the future that make economies based on resource extraction seem inevitable, and thereby preclude other forms of land use and domestic economies viable futures. See Westman, "Social Impact Assessment and the Anthropology of the Future in Canada's Tar Sands."

33. MGP, vol. 4, sec. 5, 19.

34. Beck notes for example that many of the hazards that pose a risk to societies today include items such as toxins or nuclear threats that escape sensory perceptions and can be identified through physical and chemical formulas. Ulrich Beck, *The Risk Society: Towards a New Modernity* (London: Sage Publications, 1992).

35. Ibid., 24.

36. Ibid., 26.

37. Northern Gas Project Secretariat, *Joint Review Panel for the Mackenzie Gas Project—Public Hearing Transcripts*, 1618.

38. Canadian legal jurisprudence has increasingly pointed to higher standards for the ways in which governments and industry consider Aboriginal land and treaty rights in development decisions. See, for example, *R. v. Sparrow*, *Haida Nation v. British Columbia*, *Taku River Tlingit First Nation v. British Columbia*, *Mikisew Cree First Nation v. Canada*, and *Tsilhqot'in Nation v. British Columbia*.

39. Idle No More is an ongoing grassroots movement that was sparked by federal omnibus Bill C45, which, among other provisions, saw the deregulation of a number of waterways and weakened environmental protection laws. Idle No More calls for the federal government to honor the treaties and for a restoration of indigenous jurisdiction over their territories. Idle No More is especially critical of large-scale resource extraction, which, they point out, has provided very little benefit for indigenous communities while simultaneously producing significant environmental degradation.

40. Except see Carole Blackburn, "Searching for Guarantees in the Midst of Uncertainty: Negotiating Aboriginal Rights and Title in British Columbia," *American Anthropologist* 107, no. 4 (2005): 586.

41. Author interview, April 10, 2006.

Bibliography

Abel, Kerry. *Drum Songs: Glimpses of Dene History*. Montreal: McGill-Queen's University Press, 1993.

Asch, Michael. "The Dene Economy." In *Dene Nation: The Colony Within*, edited by Mel Watkins, 47–61. Toronto: University of Toronto Press, 1977.

Basso, Ellen. "The Enemy of Every Tribe: Bushman Images in Northern Athapaskan Narratives." *American Ethnologist* 5, no. 4 (1978): 690–709.

Bavington, Dean. *Managed Annihilation: An Unnatural History of the Newfoundland Cod Collapse*. Vancouver: University of British Columbia Press, 2010.

———. "Of Fish and People: Managerial Ecology in Newfoundland and Labrador Cod Fisheries." PhD dissertation, Wilfred Laurier University, 2005.

Beck, Ulrich. *The Risk Society: Towards a New Modernity*. London: Sage Publications, 1992.

Blondin, George. *When the World Was New: Stories of the Sahtu Dene*. Yellowknife: Outcrop, The Northern Publishers, 1990.

———. *Yamoria: The Law Maker*. Edmonton: NeWest Publishers, 1997.

Clammer, John. "Decolonizing the Mind: Schwimmer, Habermas and the Anthropology of Post-Colonialism." *Anthropologica* 50, no. 1 (2008): 157–168.

Dokis, Carly. *Where the Rivers Meet: Pipelines, Politics, and Participatory Management in the Sahtu Region Northwest Territories*. Vancouver: University of British Columbia Press, 2015.

———. "Modern Day Treaties: 'Development', Politics, and the Corporatization of Land in the Sahtu Dene and Métis Comprehensive Land Claim Agreement." *Geography Research Forum* 30 (2010): 32–49.

Gagne, Natacha. "The Link Between an Anthropologist and His Subject: Eric Schwimmer and (De)colonization Processes." *Anthropologica* 50, no. 1 (2008): 13–22.

Goulet, Jean Guy. *Ways of Knowing: Experience, Knowledge, and Power Among the Dene Tha*. Lincoln: University of Nebraska Press, 1998.

Mackenzie Gas Project. *Environmental Impact Statement for the Mackenzie Gas Project*. 8 vols. Submitted by Imperial Oil on behalf of the Mackenzie Gas Project Proponents to the National Energy Board (NEB), 2004.

Merchant, Carolyn. *The Death of Nature: Women, Ecology, and the Scientific Revolution*. New York: Harper & Row, 1980.

Mills, Antonia. "The Meaningful Universe: Intersecting Forces in Beaver Indian Cosmology and Causality." *Culture* 6, no. 2 (1986): 81–91.

Nadasdy, Paul. "The Anti-Politics of TEK: The Instutionalization of Co-Management Discourse and Practice." *Anthropologica* 47, no. 2 (2005): 215–232.

Ridington, Robin. *Trail to Heaven: Knowledge and Narrative in a Northern Native Community*. Iowa City: University of Iowa Press, 1988.

Rushforth, Scott. "Legitimation of Beliefs in a Hunter-Gatherer Society: More About Bearlake Athapaskan Knowledge and Authority." *American Ethnologist* 21 (1992): 486.

———. "Country Food." In *Dene Nation: The Colony Within*, edited by Mel Watkins, 32–46. Toronto: University of Toronto Press, 1977.

Taniton, Raymond, and Mindy Willet. *At the Heart of It: Dene dzó t'árá*. Markham, ON: Fifth House/Fizhenry & Whiteside, 2011.

Westman, Clint. "Social Impact Assessment and the Anthropology of the Future in Canada's Tar Sands." *Human Organization* 72, no. 2 (2013).

Cases Cited

Haida Nation v. British Columbia (Minister of Forests), [2004] 3 S.C.R. 511, 2004 SCC 73.

Mikisew Cree First Nation v. Canada (Minister of Canadian Heritage), [2005] 3 S.C.R. 388, 2005 SCC 69.

R v. Sparrow [1990] 1 S.C.R. 1075.

Taku River Tlingit First Nation v. British Columbia (Project Assessment Director), [2004] 3 S.C.R. 550, 204 SCC 74.

Tsilhqot'in Nation v. British Columbia, 2014 SCC 44.

10

Writing in the Anthropocene from the Global North to the Global South

Mary Shelley's *Frankenstein* and Richard Powers's *The Echo Maker*

kyndra turner / INDEPENDENT SCHOLAR

AS WE PROCEED INTO THE TWENTY-FIRST CENTURY, SCIENTISTS AGREE that the ominous effects of rapid climate change, exemplified by recent devastating typhoons and rising sea levels, illustrate that "climate," as it is defined by atmospheric chemists and geographers, is disproportionately affecting economically vulnerable communities and ecosystems in both the global North and South. For example, communities in the North are disproportionately bearing the costs of the planet's warming climate as melting sea ice and receding polar glaciers affect populations of marine mammals and the livelihoods of the people who depend on them. These relationships between sources and impacts across time and space suggest that we are indeed in what atmospheric chemist Paul Crutzen and biologist Eugene Stoermer refer to as the Anthropocene—an age in which global patterns of consumption, economic development, and migration are increasing the vulnerability to environmental disaster and risk of not only humans but all life on the planet. In response, anthropologist and ecocritic Debra Bird Rose recently called for new forms of "writing in the Anthropocene" that would be "capable of shaking up our culture, and awakening [humans] to new and more enlivened understandings of the world, our place in it, and the situated connectivities that bind us into multi-species communities."[1] Rapidly changing climate demands an immediate response and, as a result, calls upon writers and academics to contribute to helping the public understand why behavioral change on the part of humans will be necessary. In this chapter, I respond to Rose's call for "writing in the Anthropocene" by engaging with the iconic North in Mary Shelley's canonical nineteenth-century novel, *Frankenstein*, and Richard Powers's contemporary novel *The Echo Maker* in order to provide models for what I call "readings in the Anthropocene," or rereadings of classic and contemporary texts that offer insights to those interested in changing human behavior and suggesting alternative practices of inhabiting the material world at the end of the twentieth century.

Focusing on the parts of the novel that are set in the North, I read Shelley's *Frankenstein* (1818) for what it suggests about the historical role that exploration, science, and nature may be playing in climatic, economic, and human and nonhuman migratory patterns that can be observed in the Anthropocene. Shelley would not have known about today's rapidly melting glaciers and rising sea levels, but she was aware of the increasing unrest in Europe due to harvest failure, riots, starvation, and a global cholera epidemic caused by climate change. In the spring of 1816—the same year Shelley started writing *Frankenstein*—worldwide temperatures dropped and patterns of rainfall changed dramatically. The stormy weather that Shelley refers to in the preface of *Frankenstein* as a "cold and rainy" season was in fact a three-year climate crisis caused by the 1815 volcanic eruption of Mount Tambora on the island of Sumbawa in what was then the Dutch East Indies, now Indonesia.[2] Environmental historian Gillen D'Arcy Wood points out that "one of the paradoxical effects of a major tropical eruption is that while the planet in general is cooled by the blanket of volcanic dust that drifts from the equator to the poles, the Arctic itself is drastically warmed owing to changes in wind circulation and north Atlantic ocean currents."[3] As the Arctic began to melt, the British Admiralty began planning a costly and ultimately unsuccessful fifty-year-long campaign to chart shorter sea routes linking Europe via the eastern North American coast to Asia. The British could not have known then, of course, that the climatic effects of Tambora would only last three years: "The Arctic refroze just in time for the arrival of Britain's first polar expedition under Captain John Ross in

1818. Years of fruitless icebound sallies into the polar seas culminated in the tragic Franklin expedition of the 1840's, when all hands were lost, and the heroic age of British Arctic exploration came to an end."[4] It is not a surprise, then, that the opening scenes of *Frankenstein* include Captain Walton's voyage through the frozen North in search of a route to the Pacific and his encounter with the Monster and its creator. Thus, the novel provides a literal record into a two-hundred-year-old event illustrating regional environmental disaster with global effects.

Juxtaposing Shelley's canonical *Frankenstein* with American National Book Award–winning novelist Richard Powers's *The Echo Maker* in relation to social and environmental issues in the global North shows how two hundred years of colonization and economic activities caused by humans result in a climatic effect that is similar to the most destructive volcanic eruption on Earth in thousands of years. A close reading of *The Echo Maker* reveals how contemporary American authors are addressing such topics as neuroscience and the arctic migration of birds to address the global North's technological power of industrial capitalism and the devastating effects of oil extraction and climate change. Powers's novel appears uncanny for anticipating some of the arguments surrounding the much-disputed proposed Keystone XL pipeline from Canada to the United States—a significant debate within the North American environmental movement. The connections that Powers and Shelley make between patterns of consumption and government and private investment at the expense of human and nonhuman bodies require readers to rethink the global North's mainstream values, economic priorities, and growing exploitative mentality. Both novels, I contend, offer insights into understanding how we can use finite resources in ways that either promote justice or enclose critical resources for the benefit of only a few people.

Polar Exploration and Scientific Enterprise in *Frankenstein*

Captain Walton's polar exploration quest for the magnetic secrets of the North Pole and its potential for travel passages for global commerce illustrates the common nineteenth-century perception of nature as an uncontested resource for human development. The polar travel narratives of Captain James Cook, Sir John Barrow, and Constantine John Phipps catalyzed Shelley's apprehension of capitalistic expeditions used to dominate nonhuman natures and cultures. For example, in Captain Walton's first letter to his sister, Margaret Saville, he asserts, "you cannot contest the inestimable benefit which I shall confer on all mankind to the last generation, by discovering a passage near the pole to those countries."[5] Captain Walton elaborates that only by undertaking such a voyage "can the secret of the magnet" be attained.[6] Here Captain Walton verbalizes his desire to dominate nature to serve man's private ends. He believes that such a discovery will give humankind the authority and power to command the elements and passageways in even the remotest parts of the globe, which will expand the British Empire and its global power. Thus, *Frankenstein* speaks to the desire of the male explorer-scientist to dominate and control nature.

Being the daughter of two of Europe's most respected intellectuals, Mary Wollstonecraft and William Godwin, Shelley was also well aware of the scientific research of Galvani, Humphry Davy, and Benjamin Franklin. However, Shelley did not see all scientific exploration as dangerous and self-serving. For example, Victor celebrates the operations of nature and science when observing lightning destroy an old tree: "I remained, while the storm lasted, watching its progress with curiosity and delight. . . . I eagerly inquired of my father the nature of thunder and lightning."[7] Yet one can also see Shelley's critique of scientific practices in which humans attempt to control or change nature, such as when Victor manipulates electricity to create a new life-form for his own self-serving purposes, which leads to the death of five innocent people: "Life and death appeared to me ideal bounds, which I should first break through, and pour a torrent of light into our dark world. A new species would bless me as its creator and source; many happy and excellent natures would owe their being to me."[8] This passage not only presents nature as passive matter that can and should be dominated and manipulated by men in order to satisfy their desire but also demonstrates a change in cultural values, attitudes, and human behavior. As Western culture became increasingly mechanized in the 1600s, the image of a living female earth gave way to a mechanistic worldview. Such a self-serving and patriarchal view of nature can be seen as the result of the Scientific

Revolution, and it is clear that Shelley, in her fiction, is arguing for something very much along these lines.

Captain Walton's pursuit to exploit nature's resources for commercial profit and control is analogous to Victor Frankenstein's destructive and narcissistic scientific exploration. The fact that Walton finds a mirror image of himself in Frankenstein—"I have longed for a friend; I have sought one who would sympathize with and love me. Behold, on these desert seas I have found such a one"—suggests that Shelley saw both forms of exploration as destructive processes driven by male egotism.[9] Both Walton and Frankenstein are products of the Scientific Revolution and have been taught to see nature as something separate from themselves, as independent passive matter that can and should be controlled and manipulated. For example, Waldman, Victor Frankenstein's professor at the University of Ingolstadt, explains to Frankenstein that philosophers (not yet "scientists" since the word *scientist* was not coined until 1833)[10] "ascend into the heavens; they have discovered how the blood circulates, and the nature of the air we breathe. They have acquired new and almost unlimited powers; they can command the thunders of heaven, mimic the earthquake, and even mock the invisible world with its own shadows."[11] Waldman encourages Frankenstein to "penetrate into the recesses of nature, and shew how she works in her hiding places."[12] This metaphor presents nature as a passive female that can be penetrated by men in order to satisfy their desire. It also provides a cultural critique about past and present human practices, such as global commerce and its effects on ecosystems.

Captain Walton and Victor Frankenstein illustrate that Shelley is musing about the ways in which the human species engages in behaviors that lead to monstrosity by failing to account for negative social, environmental, nationalistic, and economic forces, and thus the novel anticipates many of the arguments about human behaviors leading to the Anthropocene. Captain Walton's scientific theories of "the secret of the magnet" and "passage near the pole," in which he "[studied] day and night" in "good Uncle Thomas's library,"[13] resembles Frankenstein's scientific quest "into the recesses of nature" to "shew how she works in her hiding places."[14] As a result, in ecocritic Joni Adamson's view, Shelley's novel is revealed to be a "critique of science as a discourse that authorized colonial activities on a scale that today can be recognized as having altered planetary systems."[15] Thus, what I have called "readings in the Anthropocene" are absolutely critical to discussions about large- and small-scale climatic, economic, and human and nonhuman migratory patterns that have changed planetary biogeochemical processes and patterns. These new readings draw audiences into discussions of both historical texts and events and contemporary concerns that encourage us to recognize and question normative narratives that underpin our current fixation on meeting nonrenewable resource shortages, such as oil, with more finite resources and hazardous drilling, and expands our vision of the consequences of our behaviors.

Commercial Profit and Control in the Arctic North and the Anthropocene

Captain Walton's arctic expedition echoes the current race among northern nations to privatize natural resources and shipping routes in the Arctic due to the extension of the continental shelf. According to *National Geographic News* journalist John Roach, the arctic sea ice has receded about 40 percent since 1979 as a result of global climate change.[16] The disappearing sea ice could open the way to exploit a bounty of oil, gas, minerals, and fish once protected by their inaccessibility. Furthermore, arctic countries such as Russia, Canada, Denmark, Finland, Sweden, Norway, Iceland, and the United States, as well as six international organizations representing indigenous peoples and near-arctic countries such as China, recognize the potential mineral riches of the region. For example, in 2004, in a race to claim and develop the arctic North, Denmark announced a $25 million project to scientifically prove that the seabed beneath the North Pole is a natural extension of Greenland's seabed. In 2007, Russia joined the race by staking a symbolic claim to secure potential transarctic shipping routes and the region's natural resources by planting its national flag on the North Pole seafloor. By 2013, Canada had already spent over $200 million on scientific research in an effort to assert its sovereignty in the resource-rich Arctic and submitted an official claim to the United Nations Commission on the Limits of the Continental Shelf. Thus, by setting *Frankenstein* into the context of the extension of the continental shelf and the race among northern nations for potential natural resources and transarctic shipping routes one can see that "writing in the Anthropocene" as "cultural critique" is not a trend that occurs only in contemporary literature but one that writers like Shelley anticipated and began creating as far back as the nineteenth century.

Moreover, northern nations have begun competing over the economic potential of shipping routes through the Northwest Passage, across the Canadian archipelago, as well as along the Northeast Passage, north of Siberia, where melting ice may make it possible to establish an all-season trade route. Unlike the climatic impacts of Mount Tambora, which lasted only three years, these new transarctic shipping routes will not be freezing over again and as a result are becoming increasingly popular. These new routes can be "more than a third quicker for some shipments that now pass through the Suez or Panama canals."[17] Even though the shipping industry goes largely ignored by those outside of it, trade carried by sea has grown fourfold since 1970 and is still growing. In 2011, the United States' commercial ports took in international goods worth $1.73 trillion—eighty times the values of all U.S. trade in 1960.[18] Currently, ocean carriers move 90 percent of all goods. The quantity of ships using seasonal northern glacial passageways is also increasing. In 2010, only four ships sought out the Northern Sea Route, a route that crosses the Arctic Sea along Russia's northern coast. In 2011, thirty-four ships used the same crossing, and in 2012, the year of the lowest recorded arctic sea ice coverage, forty-six industrial ships crossed. In 2013, seventy-one ships navigated the arctic shipping route. Due to climate-induced arctic ice loss, by 2050 even ships not equipped with ice-breaking hulls will be able to navigate arctic shipping routes.[19] Growing shipping traffic in arctic waters also means a higher chance of acoustic pollution, greenhouse gas emissions, invasive species, oil spills, and ship strikes against whales and other marine life. By making abstract, often intangible global patterns and concepts accessible, new readings of classic literature can reveal the consequences for the biosphere of continued human activities that do not account for their ecological impacts.

The current rush for political control over the fabled Northwest Passage between the Atlantic and Pacific Oceans—and the natural resources that lie beneath it—has important economic, environmental, and governance implications for not only northern nations but also communities in the North. The melting arctic sea ice has put the Inuit people at the center of the debate over the future of the arctic region, and whether commercial and capitalistic interests can trump the needs of the environment and northern peoples. Inuit communities across Canada, Greenland, Alaska, and Russia rely on marine resources to support their subsistence way of life and cultural traditions. Aqqaluk Lynge, chairman of the Inuit Circumpolar Council, which represents more than 160,000 Inuit people, reminds political leaders "that you cannot use the Arctic as a laboratory. It's not a laboratory. The Arctic Ocean is not the last frontier. It's our home. People have to remember that people live there."[20] The actions of Victor Frankenstein and Captain Walton illustrate how the creation of "wealth" rests on the control and extraction of materials from the environment. The enclosure of arctic resources for the wealthy represents the undemocratic relationship among species and among human groups. The tangled narratives of discoverer, scientist, and creature in the Arctic shed light on contemporary concerns and challenge readers to rethink and reconceptualize normative narratives in order to transform human behavior and our place in the world.

Human Behavior in *The Echo Maker* and the Anthropocene

Similar to Shelley, Powers explores the cognitive process and behavior of human and nonhuman species. *The Echo Maker* takes place in rural Kearney, Nebraska, where Mark Schluter, a twenty-seven-year-old meatpacker, mysteriously flips his truck while driving at night through the sandhill crane migration range. When Mark wakes after a fourteen-day coma, he believes his sister, Karin, to be an imposter along with his dog, Blacky, and his house, "The HomeStar." Mark's condition is identified as Capgras syndrome, a rare neurological disorder where sufferers fail to recognize those people and possessions closest to them. Mark's circumstances are so extraordinary that Dr. Gerald Weber, a famous cognitive neurologist, agrees to examine him. In an interview with Alec Michod, Powers explains how Mark's brain injury and crane intelligence gave him "a way to open up the story to all kinds of neurological and ecological traces."[21] For example, Powers elaborates, "so here we are, sharing the planet with these creatures who are weirdly intelligent, smart in an alien way that we're not quite smart enough to see. And yet, the core parts of their brains are still contained in ours."[22] Over the course of the novel, Powers uses the brains and the intelligence of both sandhill cranes and humans to raise questions about mainstream, self-serving behaviors that are putting human and nonhuman species at risk in the Anthropocene.

For example, Dr. Weber's mercenary attempt to control Mark's brain through artificial means and Victor Frankenstein's self-indulgent attempt to "[acquire] unlimited powers" could be read as metaphors for the ways in which human activities today are trumping the needs of nonhuman species and people.[23] Dr. Weber, neuroscientist and famous author of popular neurological case studies, arrives in Kearney, Nebraska, after receiving a desperate plea from Karin to examine her brother, Mark. Over the years Dr. Weber's research has given him publicity, money, and his made-up alter ego, "Famous Gerald." As a result, Dr. Weber fails to see Mark's brain as something more than independent passive matter for him to control and manipulate. Powers challenges readers to see that Mark's brain is not an isolated agent or, to use environmental scientist Brad Allenby's words, "a design space" where we use cognitive science and emerging technologies—nanotechnology, biotechnology, robotics, information and communication technology—to design our internal space for our benefit. However, the results are unpredictable. Weber dishearteningly asserts, "Psychopharmacology: hit or miss, hard to tune, ripe with side effects, symptom-masking, and once begun, difficult to tail off of."[24] In the end, Dr. Weber does not "cure" or "redesign" Mark: "When Karin Schulter enters her brother's room, the man who has been denying her is gone. In his place, a Mark she has never seen sits in a chair in striped pajamas."[25] Weber's failed attempt to use science and technology to quickly fix Mark's behavior suggests that redesigning the human with the use of science and technology is not the answer to changing human behavior in the Anthropocene. Powers is asking his readers to think about the disastrous consequences of self-serving enterprises that put others at risk and the possibility of "ordinary" citizens to change their behaviors.

Human and Nonhuman Migration in the Anthropocene

Captain Walton, Dr. Frankenstein, and the Monster's complicated journey north parallel the migratory flight north of the cranes in *The Echo Maker*. Powers begins *The Echo Maker* and each of its next four sections with cranes migrating north across the length of the hemisphere, with a migratory stopover on the Platte River in the flat wetlands of Nebraska. The endangered sandhill cranes, known to indigenous people as "the echo makers" because of their sonorous calls, "push up from New Mexico, Texas, and Mexico, hundreds of miles each day, with thousands more ahead before they reach their remembered nests."[26] The crane's migration puts human migration from place to place in perspective. Unlike migratory animals, humans often do not leave a place with its integrity intact but rather they change the ecosystem to such an extent that the species living there cannot endure. Powers shows clearly in the novel that the sandhill cranes are so crowded into smaller and smaller spaces on the Platte as humans expand their own communities that the birds are becoming more and more diseased and disoriented. Yet Karin's work at the Buffalo County Crane Refuge also illustrates that one can be place-based locally while cultivating deep concern for a broader spectrum of global movement. Karin, who knows almost nothing about cranes, reaches out to the community by volunteering at the sandhill crane reserve. As Karin explains to fictional developer and con man Robert Karsh, about her work at the Crane Refuge: "What can I say? It's the most fulfilling work I've ever done. Bigger than myself? How about bigger than anyone. I'm working through some papers . . . Did you know that we've changed that river more in one hundred years than in all the ten thousand years prior?"[27] The sandhill cranes expand the reader's sense of place and civic responsibility by illustrating a biospheric perspective to local ecologies and communities.

In contrast to bird migration, concepts such as Manifest Destiny in the nineteenth century propelled human migration in the United States from east to west. Incentives for migration such as the Homestead Act, the California Gold Rush, and the mysterious wilderness of the Western frontier continued to increase westward expansion in America throughout the nineteenth century. The setting of the novel, Kearney, Nebraska, lies on or near the great historical east-to-west routes: the Oregon Trail, Mormon Trail, Pawnee buffalo trail, Pony Express, transcontinental railroad, Lincoln Highway, Interstate 80.[28] Here the east–west human migration corridor crosses the ancient north–south migration route of the continent's birds. On one route, cranes fly from northern Mexico to the Platte River basin in Nebraska, where sandhill cranes rest and feed as they prepare for the arduous journey across vast tundra in Canada, Alaska, and Siberia. As the sandhill cranes migrate through the Central Flyway, humans continue to move east to west, not only reinforcing the great routes of commerce and cultural exchange but also shaping and transforming the environment through accelerated unprecedented human activities indicated by industrial agriculture, modification of water and sediment, as well as urbanization and industrialization, which have played a role in catalyzing the transition of the Holocene into the Anthropocene.

Rights and Access to Communal Spaces and Resources

The crane's shrinking habitat is a symptom of human ego and behavior that is putting the livelihoods of individuals and groups at risk by ignoring social costs and environmental impacts. Each year the birds at the Platte River look more and more spectacular. Each year it seems as if there are more and more cranes, "coming into town in packs. [Bunches] of them [landing] on the roof of the McDonald's."[29] However, what appears as an increase in crane population is actually only a symptom of their declining wetland habitat. In the novel, crane conservationist and ascetic environmental activist Daniel Riegel explains to Karin how the cranes "used to roost along the whole Big Bend: a hundred and twenty miles or more. They're down to sixty, and shrinking. The same number of birds crammed into half the space."[30] The development of strip malls, housing complexes (such as the one Mark lives in at River Run Estates), and industrial agriculture have diminished the cranes' habitat. The cranes are spectacular "precisely because the river had drained away beneath them, concentrating them in a few remaining heavens."[31] The plight of the sandhill crane is further complicated by local development of water resources.

Interestingly, the cranes in *The Echo Maker* illustrate complex dilemmas on the local and global level that characterize the Anthropocene. The illustration of resource extraction anchors the novel in real-world issues concerning the global freshwater crisis and regional access to clean water supplies. Powers shows that the challenge we face now is how to effectively conserve, manage, and distribute the water we have. For the sandhill cranes, the problem is access to water and nesting ranges. Riegel alarmingly explains that if "the river slows; the trees and vegetation fill in. The trees spook the cranes. They need the flats—someplace to roost where nothing can sneak up. . . . This is their only safe stopover. No other spot in the center of the continent they can use."[32] However, not everyone in the community believes in placing the birds' sanctuary ahead of the comfort and welfare of the people. Water is the difference between wetlands and irrigated fields, between profit and loss. The crane refuge, where Daniel works, is attacked from all sides. A local farmer verbally assaults Daniel personally: "'You have any idea how much damage those birds do? . . . It took Americans hundreds of years to turn this swampland into beautiful farms. And you people want to turn it back into swamps again.'"[33] Powers scrutinizes the difficult and delicate balance between land conservation and agriculture, and the importance of taking an active role in one's social and environmental surroundings. He offers a more nuanced and complex understanding of the meaning of land management and water conservation, and calls for both human and nonhuman rights to natural resources.

Transnational Monsters

Through the trajectory of the cranes' migration, Powers pushes the reader's focus from the local community of Kearney across international borders to the global North's devastating effects of industrial production on the natural world. This wider hemispheric perspective highlights some of the modern transnational monsters we are facing today as we proceed into the twenty-first century. For example, TransCanada's $7 billion pipeline raises many environmental concerns. The Keystone XL pipeline would funnel tar sands more than seventeen hundred miles from northern Alberta, Canada, to refineries along the Gulf of Mexico, crossing international borders. Radical environmentalist activist and writer Aric McBay explains that after the tar sands have been extracted from living forests and soils on the surface, "water drained by nearby rivers is used to wash the bitumen out of the sand—several volumes of water are used for every volume of oil—leaving a toxic water-oil by-product that kills fish, birds, and indigenous people living in the area."[34] McBay elaborates, "if you simply hated the land and wanted to destroy it, you would be hardpressed to find a more vicious way of doing it."[35] The migrating cranes in *The Echo Maker* reveal an expansive understanding of the relationship between industrialism, human bodies, and the more-than-human world. Similarly, Captain Walton's polar exploration for the magnetic secrets of the North Pole and his contact with Victor Frankenstein and the Monster become a metaphor for the ways in which human activities today are impacting the world's most vulnerable nonhuman species and people in the global North. Thus, placing regional environmental and social concerns, such as land management, water conservation, and human and nonhuman migration patterns, into the context of transnational disputes over the Keystone XL pipeline and arctic minerals and passageways challenges consumer consumption and growth and, as a result, insists that behavioral change among humans will be necessary—not at some ambiguous time in the future but now.

The Keystone XL pipeline also brings the reader's attention back to the crossroad setting of the novel where north–south crane migration patterns intersect east–west economic routes of human consumption and growth. Debate over the proposed route of the pipeline has drawn international attention focused largely on Nebraska. The planned route for the pipeline would run through Nebraska, crossing the ecosensitive Sandhills and the vast Ogallala Aquifer, a vital source of drinking water for the Great Plains. In a recent independent study, environmental engineer John Stansbury estimated that a major spill from the proposed Keystone XL where the pipeline crosses the Platte "could threaten the drinking water of hundreds of thousands of people as far south as Kansas City, Missouri."[36] Such a leak would subject local Sandhill residents as well as communities and ecosystems living downstream to toxic drinking water, carcinogens, and polluted agricultural and recreational areas. Powers demonstrates how resource extraction is inseparable to the destruction and displacement of human and nonhuman bodies. Thus, the novel is a "reading in the Anthropocene" that enhances recognition of America's current fixation on meeting nonrenewable resource shortages, such as oil, with more finite resources and high-tech pipelines, and expands our vision of the consequences of our behaviors and what constitutes a sustainable environment.

Powers's open ending leaves the novel hanging on a dystopian note, making it clear that we are indeed at the tipping point of our environmental crisis: "The outcome of owls will orchestrate the night, millions of years after people work their own end. Nothing will miss us."[37] Left with no definite answers or solutions, readers are forced to reexamine their choices and actions in order to better understand one another's social and environmental connections. Karin's volunteer work at the Buffalo County Crane Refuge reminds readers that "ordinary people" of all ages, races, and disciplines (such as Karin and the Keystone pipeline protestors) have the power to evolve and can come together to create transnational coalitions and an international environmental response large enough to make a difference.

By connecting these two texts, readers in the Anthropocene can frame narratives to mobilize change and/or build coalitions to confront the damaging effects of resource extraction and climate change. Both texts urge "ordinary" citizens to consider how they might play a role in addressing both local and global social and environmental justice challenges. Shelley's Dr. Frankenstein illustrates how the human species might continue engaging in behaviors that lead to monstrosity by failing to account for the negative environmental, political, and cultural forces at work in the Anthropocene. *The Echo Maker*'s Karin represents an "ordinary" citizen becoming educated about and getting involved in debates over environmental and social justice challenges that are global in scale. The global North's "mainstream" values, priorities, and attitudes are leading to oil extraction in vulnerable ecosystems and other actions that are killing the planet. These novels call upon people both inside and outside the academy to rethink mainstream perspectives in order to transform human behavior and reconceptualize "the human's" place in the world. Both Shelley's classic nineteenth-century novel *Frankenstein* and Powers's contemporary novel *The Echo Maker* transcend hemispheres, oceans, and disciplines, and are exemplary of literature of the Anthropocene. That these texts center on the geographic North as a way to make this argument is telling: the North's concerns have transnational import. It is no wonder that the world's environmental future hinges on dramas being played out in these spaces.

Endnotes

1. Debra Bird Rose, "Introduction: Writing in the Anthropocene," *Australian Humanities Review* 49 (2009): 87.

2. Mary Shelley, *Frankenstein, or the Modern Prometheus*, ed. J. Paul. Hunter (New York: W.W. Norton & Company, 1996), 6.

3. Gillen D'Arcy Wood, "The Volcano That Changed the Course of History," *Slate Magazine*, April 9, 2014, accessed June 5, 2014, http://www.slate.com/articles/health_and_science/science/2014/04/tambora_eruption_caused_the_year_without_a_summer_cholera_opium_famine_and.html.

4. Ibid.

5. Shelley, *Frankenstein*, 8.

6. Ibid.

7. Ibid., 23.

8. Ibid., 32.

9. Ibid., 147.

10. Tim Fulford, Debbie Lee, and Peter J. Kitson, *Literature, Science and Exploration in the Romantic Era: Bodies of Knowledge* (Cambridge: Cambridge University Press, 2004), 170.

11. Shelley, *Frankenstein*, 28.

12. Ibid., 29.

13. Ibid., 8.

14. Ibid., 29.

15. Joni Adamson, "Humanities," in *Keywords for Environmental Studies*, ed. Joni Adamson, William Gleason, and David N. Pellow (New York: New York University Press, 2016).

16. John Roach, "As Arctic Ice Melts, Rush Is on for Shipping Lanes, More," *National Geographic News*, February 25, 2005, accessed January 2, 2014, http://news.nationalgeographic.com/news/2005/02/0225_050225_arctic_landrush.html.

17. Ibid.

18. Rose George, *Ninety Percent of Everything: Inside Shipping, the Invisible Industry that Puts Clothes on Your Back, Gas in Your Car, and Food on Your Plate* (New York: Metropolitan Books, 2013), 3.

19. Laurence C. Smith and Scott R. Stephenson, "New Trans-Arctic Shipping Routes Navigable by Midcentury," *Proceedings of the National Academy of Sciences* 110, no. 13 (2013): 4871–4872.

20. James Brooke, "As North Pole Ice Melts, More Ships Take Arctic Shortcut," *Voice of America*, October 4, 2013, accessed February 23, 2014, http://www.voanews.com/content/as-north-pole-ice-melts-more-ships-take-arctic-shortcut/1763072.html.

21. Alec Michod, "Interview with Richard Powers," *The Believer*, February 2007, accessed November 3, 2011, http://www.believermag.com/issues/200702/?read=interview_powers.

22. Ibid.

23. Shelley, *Frankenstein*, 29.

24. Richard Powers, *The Echo Maker* (New York: Farrar, Straus and Giroux, 2006), 311.

25. Ibid., 443.

26. Ibid., 4.

27. Ibid., 329.

28. Michod, "Interview with Richard Powers."

29. Powers, *The Echo Maker*, 16.

30. Ibid., 57.

31. Ibid., 346.

32. Ibid., 57.

33. Ibid., 264.

34. Aric McBay, "Civilization and Other Hazards," in *Deep Green Resistance: Strategy to Save the Planet*, ed. Aric McBay, Lierre Keith, and Derrick Jensen (New York: Seven Stories, 2011), 32.

35. Ibid.

36. Sandra Postel, "Keystone XL, Clean Water and Democracy," *National Geographic Magazine* (November 2011), accessed November 2, 2011, http://newswatch.nationalgeographic.com/2011/11/16/ keystone-xl-clean-water-and-democracy.

37. Powers, *The Echo Maker*, 443.

Bibliography

Adamson, Joni. "Humanities." In *Keywords for Environmental Studies*, edited by Joni Adamson, William Gleason, and David N. Pellow. New York: New York University Press, 2015.

Brooke, James. “As North Pole Ice Melts, More Ships Take Arctic Shortcut.” *Voice of America*, October 4, 2013, http://www.voanews.com/content/as-north-pole-ice-melts-more-ships-take-arctic-shortcut/1763072.html.

Fulford, Tim, Debbie Lee, and Peter J. Kitson. *Literature, Science and Exploration in the Romantic Era: Bodies of Knowledge*. Cambridge: Cambridge University Press, 2004.

McBay, Aric. “Civilization and Other Hazards.” In *Deep Green Resistance: Strategy to Save the Planet*, edited by Aric McBay, Lierre Keith, and Derrick Jensen, 31–59. New York: Seven Stories, 2011.

Michod, Alec. “Interview with Richard Powers.” *The Believer*, February 2007, http://www.believermag.com/issues/200702/?read=interview_powers.

Postel, Sandra. “Keystone XL, Clean Water and Democracy.” *National Geographic Magazine*, November 16, 2011, http://newswatch.nationalgeographic.com/2011/11/16/keystone-xl-clean-water-and-democracy.

Powers, Richard. *The Echo Maker*. New York: Farrar, Straus and Giroux, 2006.

Roach, John. “As Arctic Ice Melts, Rush Is on for Shipping Lanes, More.” *National Geographic News*, February 25, 2005, http://news.nationalgeographic.com/news/2005/02/0225_050225_arctic_landrush.html.

Rose, Debra Bird. “Introduction: Writing in the Anthropocene.” *Australian Humanities Review* 49 (2009): 87.

Shelley, Mary. *Frankenstein, or the Modern Prometheus*. Edited by J. Paul Hunter. New York: W.W. Norton & Co., 1996.

Smith, Laurence C., and Scott R. Stephenson. “New Trans-Arctic Shipping Routes Navigable by Midcentury.” *Proceedings of the National Academy of Sciences* 110, no. 13 (2013): 4871–4872.

Wood, Gillen D’Arcy. “The Volcano That Changed the Course of History.” *Slate Magazine*, April 9, 2014, http://www.slate.com/articles/health_and_science/science/2014/04/tambora_eruption_caused_the_year_without_a_summer_cholera_opium_famine_and.html.

PART III

notions of north and nation

indigenous norths

11

Surveillance and the Self

Two Sami Filmmakers Explore Indigenous and Personal Sovereignty across Sápmi "Borderlands"

cheryl j. fish / BOROUGH OF MANHATTAN COMMUNITY COLLEGE, CITY UNIVERSITY OF NEW YORK

Introduction

The movie *Pathfinder* (*Ofelaš*) by Nils Gaup, based on ancient Sami legend, put Sami film on the world's stage with its 1987 nomination for an Academy Award for best foreign film. Since then, contemporary Sami film might be said to unite the technology of film with stories of an ancient people from the North who are indigenous and not quite "white." A relatively small group (around fifty thousand to seventy thousand people), the Sami, known historically as Lapps, are a northern people whose homeland transverses four countries—Norway, Sweden, Finland, and the Kola Peninsula of Russia. Sami identities are different from and yet intersect with other northern populations and indigenous groups. Sami filmmakers employ complex strategies in their documentary and fictional films: they engage with the role of place, culture, history, and political and environmental activism to create narratives of their lives both above and below the Arctic Circle.[1] Agency and resistance to colonial and national narratives has been a dominant theme even as it intersects with personal stories that abound with irony, humor, and self-surveillance.

Here I must briefly discuss the *yoik*, a form of Sami song that includes scale and vocalizations with which most persons in the West are unfamiliar, as it has been representative of the encroachment and abuse that the Sami people have suffered at the hands of outsiders. It has also been the traditional Sami way of remembering, aurally communicating innermost feelings, as well as drawing on metaphorical imagery that transcends time, persons, and landscapes. Thus it seems a fitting metaphor for film, a medium that conveys conscious and unconscious access to time, place, subjectivity, and myth.[2] In fact, in one of the films on which this essay focuses, Liselotte Wajstedt's *Sami Daughter Yoik* (*Sami Nieida Jojk*), the director claims her film functions as a *yoik*.[3]

I am interested in how emerging generations of Sami filmmakers have been framing and reinvigorating Sami experience, history, culture, and political sovereignty through film, including documentaries and fictional films, long and short, and which include English subtitles to attempt to reach a wider audience.

The film through which other Sami films signify both directly and indirectly is *Pathfinder* (*Ofelaš*), by Nils Gaup, with its depiction of the struggles of a nomadic people in a harsh northern environment that face threats from within and without, and rely on the guidance and wisdom of a shaman. The many works of Paul-Anders Simma about contemporary Sami experience, culture, and traditions, told with passion and humor, such as *Let's Dance, Minister of the State,* and *Give Us Our Skeletons,* have been shown widely in Scandinavia and abroad. Sami films have been screened in Europe and North America, at film festivals and elsewhere. In New York City, where I live, Scandinavia House and the Museum of Modern Art are two important cultural institutions that regularly sponsor Sami film and invite the directors to speak.[4] Film is a powerful medium in our increasingly visual times; it enables the artist to constitute what Elina Helander and Kaarina Kailo call "epistemological truths" for Sami: engaging with storytelling, evaluating previous activities, memorializing phenomenon, drawing on spirit and the word, and relying upon intuition.[5] The Sami draw on regional, environmental, and social concerns that coexist with transnational disputes over resources and indigenous rights. In doing so, as I conclude, they deploy what Gerald Vizenor calls "survivance": a concept combining survival and resistance, challenging dualistic notions of dominance and victimhood.

Wajstedt's *Sami Daughter Yoik* and Kalvemo's *Fear of the Boundless*

In this essay, I focus on two documentaries, Liselotte Wajstedt's *Sami Daughter Yoik* (*Sami Nieida Jojk*) (2007) and Johs Kalvemo's *Fear of the Boundless* (*Frykten for De Grenselose*) (2002).[6] They blend personal narratives with references to colonialist policies and resistance, Sami cultural identity, and the role of surveillance as the filmmakers consider their insider/outsider roles. Both involve journeys into the northern corners of Sápmi, the area within the four countries above the Arctic Circle that are Sami homelands. These two films in particular speak to me of the richness of the emerging Sami tradition that incorporates myth, history, journalistic inquiry, and personal narrative to make compelling and entertaining art that speaks to anyone with an interest in human rights, social justice, and indigenous life. And following what Susan Kollin writes in her contribution to this volume, Sami directors may be cosmopolitan artists who travel frequently, which also bears on their work.

Wajstedt's film is based on her year-and-a-half journey into her Sami heritage on her mother's side, and her own longing to reclaim her Sami identity. It is a travelogue too, as Wajstedt heads north from her adopted hometown of Stockholm to her childhood home, Kiruna, Sweden, and then her Sami spiritual home, Kautokeino, Norway. She draws the map on screen and psychically as she shifts among these locations and the motion of her car on the road. The farther north Wajstedt travels, the deeper the spell is cast to lift her suppressed Sami self. Passing trees and landscape create a sense of liminality, and the wilderness becomes increasingly magical as well as threatening. As Harald Gaski notes, Sami oral descriptions of landscape traditionally function as maps, "in which are incorporated topography, geography and information."[7] Wajstedt is crossing a threshold, in anthropologist Victor Turner's definition, to create her sense of being "neither here nor there . . . betwixt and between the positions assigned and arrayed by law, custom, convention, and ceremony."[8] *Sami Daughter Yoik* constantly negotiates these categories; the filmmaker is both participant and observer, questioning her role as storyteller too, creating an epistemology of uncertainty with which a viewer may engage. Identity can never be taken for granted or easily assumed and is partial or ambivalent. Speaking as a woman who seeks to identify as Sami as she heads to her true north illustrates interlocking historical discourses of colonial and patriarchal practices—the repression and forced assimilation of a people, and the fear of making a space for female subjectivity. They inform each other, but may not be acknowledged.[9] Previously, I coined the term *mobile subjectivity*—"a fluid and provisional epistemology and subject position contingent upon one's relationship to specific persons, incidents, ideologies, locations, time and space"—to describe subject positioning in women's travel narratives in the nineteenth century.[10] This concept also seems appropriate here. Contingent and slippery, Wajstedt's subjectivity depends on geopolitical location, state of mind, and sense of abjection. In this film, the protagonist's perspective is relative to insider/outsider status, location within/without Sami territory, and the set of beliefs that indicate a Sami or Swedish woman's ways of being.

On the other hand, Kalvemo's film presents mobile subjectivity as influenced by the outsiders who watch because they suspect him and other Sami of transgression. *Fear of the Boundless* explores the ways in which the Norwegian security police kept many Sami leaders under surveillance from the 1970s until the 1990s, as the threat of a people who are not bound by national borders and the cold war atmosphere with the Soviet Union spurred superstition and misunderstanding of the Sami people's desire for autonomy and survival. Even today, the Sami in the Kola Peninsula are threatened with extinction, "especially if language is regarded as one of the most important characteristics of an ethnic group."[11] According to Janike Kampevold Larsen and Peter Hemmersam's essay in this volume (ch. 8), the Barents Sea area and the Kola Peninsula today are marked by "mining, military activities, and survellience operations."[12] Kalvemo's film is also a journey into his own past, as the Sami radio journalist and filmmaker finds out he had long been kept under surveillance.

Assimilationist policies regarding the Sami go far back: "The assimilation policy of minorities in Norway started in the 1860s, and was seen as an issue of national security, as minorities were considered a potential danger . . . the fear of Russia also brought mistrust towards Swedish and Finnish reindeer herders coming to the Norwegian coast."[13] One way the Norwegian government dealt with this fear was to use education and social

Fig. 11.1. Filmmaker Liselotte Wajstedt wearing a Sami *gakti*, fraught with meaning in her film.

welfare as a means of swift assimilation; in Sweden, too, an attitude of "social Darwinism" toward Sami justified classifying them as inferior.[14] One can see that attitude toward the Sami in a popular novel from the early twentieth century. In *Growth of the Soil* (1917), Norwegian author Knut Hamsun portrays Sami people as physically and mentally inferior and as superstitious beggars who take advantage of a successful Norwegian homesteader, Isak, whose love and understanding of the soil, as well as his ambition and national pride, enable endearing citizenship.[15] Isak is seen as deserving of social capital while the Sami are portrayed as the nomadic interlopers of the North, a threat of disturbance that lives within.[16]

Sami Epistemologies: Breaking Verbal and Visual Silences

In both films, Sami epistemologies move to rectify imbalances in the perception and representations of Sami by outsiders whose observations have affected different generations of Sami unevenly. According to Elina Helander and Kaarina Kailo in *No Beginning, No End: The Sami Speak Up*, "it is not customary for the Sami to communicate by revealing everything. The Sami get suspicious of anyone who is too open and frank."[17] But in these films, the director/actors break the silence on both a visual and verbal level as a sign of Sami artistry and personal self-assertion; through the act of seeing and reenvisioning, a viewer identifies and comes to sympathize with the Sami protagonist's quest. In the case of Wajstedt, it is her desire to belong and fear of failure, begun with an attempt to study Sami language intensively in a school in Sápmi, only to leave tearfully. She figures herself as a "failed copy" of a true Sami, visually representing her inner struggle but contextualizing it in family and social history. She repeats throughout the film that her mother did not teach her any Sami language or expose her to her family's rich heritage while she was growing up, and the pain is expressed doubly; first, by her mother, who tells her, "I was called Lappish bastard so I had no desire to teach my children Sami," and then by Wajstedt, who feels the pain and the emptiness of such an omission.[18] These comments evoke a history of the generation who was forced to attend Swedish, Norwegian, Finnish, or Russian schools that suppressed

Sami culture and languages in enforcing assimilation policies. Next, there is the despair and longing of the younger generation, whom Wajstedt represents, to bridge the gap between modernity and tradition, between Swedish and Sami culture. They seek to become agents of transformation and live in what Gloria Anzaldúa called the "borderlands"—a place of mixing and crossing back and forth, living with multiple identities and finding a way to become comfortable but never complacent.[19] Wajstedt's background in avant-garde artistry and performance art also contributes to her sensibility in the film by relying on various postmodern techniques: collage, animation, talking bubbles from comic strips and graphic novels, music, and camera mise-en-scène indicating motion or stasis or special effect. All of these are incorporated as her persona shifts between locations and sensations and contributes to the sense of liminal identity in the film, for the protagonist and for the viewer.

Kalvemo's background as a radio journalist who claims a public space for Sami interests is apparent in the reportorial tone he assumes in his film after living in the borderland between "journalist/informer" and "person of interest" to the authorities. His film is a dialogue that takes place within that sense of clandestine objectification. By breaking the silence in his film, he exposes the institutional and state authorities that had put up a wall that could have lead to silence, fear, and withdrawal.

The Sami *Gakti* as Self-Surveillance

Sami language is one of the basic markers of ethnic identity. For Wajstedt, fashioning and wearing the Sami *gakti* (*kolt* in Swedish)—a traditional Sami costume worn for ceremonial purposes, characterized by a dominant color and decorated with plaits, embroidery, belts, or silver buttons—becomes the film's central motif. This creates a relationship between object, artist, and viewer, figuring a sense of embodiment. The iconography of the *kolt* becomes a form of self-surveillance: do the clothes fit? Does wearing it make her feel Sami? Traditional Sami clothing was strongly connected to the origins of the group, since the pattern, shape, and details provided information about the wearer's geographical origin[20]—in *Sami Daughter Yoik*, making and putting on clothes is an elaborate and self-conscious ritual. When Sami are wearing traditional clothing among non-Sami populations, they can be marked as indigenous and stand apart from other white or Scandinavian people.[21] Therefore, as the protagonist documents her experience she toys with notions of authenticity. Authentic—perhaps. But for whom? What is the difference between feeling Sami and being Sami? The Sami *gakti*, for instance, has been worn by "fake Sami" greeters at the airport in Rovaniemi in the north of Finland as a sign of exotic otherness to welcome arriving tourists from southern Europe and Asia who want to experience the North by seeing snow and visiting Santa Claus.[22] While Wajstedt does not address this particular form of mimicry in her film, the example in Rovaniemi and the Sami response to it emphasizes how wearing the *gakti* is a conscious political act by some Sami to signify authenticity and self-recognition within and outside Sami culture.

"Threats" to National Security?

The Sami filmmaker's gaze might be understood as a response to the disciplinary gaze and blunt power of the authorities to label and name adversaries. For Kalvemo, surveillance of Sami leaders who are deemed possible threats to Norwegian national security evokes Michel Foucault's panopticon, the "all-seeing" gaze that functioned as a round-the-clock surveillance machine in prisons. Its design ensured that no prisoner could ever see the "inspector," who conducted surveillance from the privileged central location. The prisoner could never know when he was being watched—mental uncertainty in itself would prove to be a crucial instrument of discipline and create an unequal gaze. Kalvemo's film claims power for Sami people and challenges the disciplinary function of nation and what he and other Sami call "ethnic surveillance."

Kalvemo is older than Wajstedt; his film evokes cold war paranoia and the post–World War II suspicion that somehow the Sápmi nation is considered unpatriotic to any country; he inserts footage in the beginning of the film to establish Sami participation to defend Finland in the Finno-Russian War and show Sami people in the Russian Arctic who joined to fight the Germans and carry supplies to the front by reindeer. The Norwegian special branch, however, believed that in the 1970s, when the Sami began to organize protests against the building of hydroelectric projects in northern Scandinavia (in Alta, Norway, and other locations) and other industrial development that affected reindeer herding areas, there was a dangerous Sami underground linked to the Nordic Sami Council's leaders. The council was formed in 1956 to promote cooperation between Sami people in different countries. Kalvemo's film includes interviews with many of the so-called radical Sami leaders who have come under surveillance for decades. One is Odd Mathis Haetta, a professor at the Sami College in Norway, who identifies himself as a pacifist. In the film he says, "it puzzled me that they thought I'd resort to armed struggle to undermine the status quo to establish a Sami state."[23] Instead, he counters, they wanted to develop their community on their own terms. Haetta confirms that Norwegian society's lack of knowledge about the Sami people lead to the chastisement of innocent people. The Sami protests in Alta were nonviolent; however, there was an explosion that blew up the Skibotn Bridge and the authorities suspected "far left Sami from Scandinavia and the Soviet Union." Because the Sami cross borders and have extensive family ties, the police interrogated and harassed many Sami families. The media at the time portrayed the Sami as militants or even terrorists. By letting the "accused" Sami speak in the film, Kalvemo attempts to set the record straight. He establishes a Sami counter-narrative based on the historical incidents and alternative claims.

Nilaas Somby, who was involved in all the Alta protests, is interviewed in the film and takes us back to the scene of the crime and discusses his motives and actions. He points to the spot on the bridge where he and a friend placed a charge: "I had to get my message across," he says. "The development was a violation of Sami rights . . . the charge wasn't enough to blow up the bridge . . . in fact it went off by accident." Somby lost an eye and arm. He was sent to jail for six months and "was made into the biggest terrorist in this part of the world at the time." Somby, who was interviewed at length in the Helsinki English newspaper *Six Degrees*, said the protests began because the "Sami people refused to accept the authorities saying that reindeer herding doesn't count (as a legitimate profession) because there's no economy in it."[24]

The infringement of nation-state borders in the eighteenth century has been replaced by logging, mining, tourism, and hydroelectric development, all of which are examples of colonial encroachment and neoliberal globalization that have affected the traditional Sami way of life. In terms of Sami women, not the focus of Kalvemo's film, these developments "resulted in the loss of both their status and visibility in their livelihood and in policies regulating it."[25] Incorporation of the Alta hydroelectric protests is central to Kalvemo's film and is a frequent signifier of Sami resistance: "the Alta affair was . . . a mega happening which accentuated all the vital aspects of the Sami situation as a people, as an ethnic minority population in Norway, and as an indigenous people in a global perspective."[26]

Kalvemo realizes that because of his movements into and out of the Soviet Union in the 1970s, for his journalism and personal connections in the city of Murmansk, he was kept under surveillance by the Norwegian special branch. He says, somewhat ironically, "I was afraid of the KGB," because the Soviets would not hesitate to use the Sami people for political purposes, and a strict policy of assimilation was imposed on the Sami during the Soviet period,[27] but he didn't realize the government of Norway feared him because of the "sensitivity" of the border and his role as one who crossed back and forth. His film is a response to such surveillance, to the suspicions of people who don't identity with nationalistic conceptions; however, he doesn't elaborate on the degree to which the Sami parliaments and other direct action protests have helped in regard to Sami and their state relations.

Wajstedt also addresses Sami political activism in her film to make the point of its significance to Sami rights and sovereignty, and of her own learning process. Her film includes scenes of Sami people marching wearing *gaktis* in Stockholm in support of the International Labor Organization's Article 169 to secure indigenous people's rights, and later shows the former Minister of Integration's press conference with Sami people stating that the Swedish government was still not willing to change regulations on fishing and hunting rights for those who do not own reindeer. The filmmaker explains how that policy affected her dad and brother, who, as non-Sami, nonreindeer owners, wouldn't be able to hunt and fish as much as they liked, but that signing the treaty would support Sami land rights. She includes an exchange with another Sami woman, more politically aware, while she attends an Easter Festival that is stressful; this young woman explains how the ILO helped the Sami in Norway and asks, "Why can't Sweden sign too?" Wajstedt concludes, "I don't even know all the things I should know about Sami people."[28] It is part of her larger crisis of doubt, and one potential conflict for those of mixed Sami and Swedish parentage and as a woman who may or may not be comfortable in a public political spotlight.

As the film progresses, Wajstedt returns to the politics of clothing—her aunt has sewn her *kolt* from handpicked fabrics and trimmings. She has worn it on several trips, first at her cousin's wedding, where she says she "can't relax in my new clothes," then to festivals where hundreds of young Sami people dance to rock bands covering Sami *yoiks* where Wajstedt struggles to find acceptance. She dons traditional Sami shoes but, she says, they "never quite fit."

Sami Mysteries and the *Yoik*

The spiritual aspect of Sami epistemology and identity comes across more deeply in Wajstedt's exporation of the Sami *yoik* and the sacred spiritual pilgrimage she makes to the "place of sacrifice," which is addressed in a powerful scene.

Wajstedt calls her movie a *yoik*—a dedication to what you wish to remember. "You can *yoik* a longing," she states in her voiceover. She travels to a music festival and interviews DJ Ante about the meaning of *yoik*ing, to replace the negative information her aunt had supplied. "It was a sin for a Christian Sami to *yoik*," and for that reason her aunt called it "howling." According to Hugh Beach, a *yoik* has its roots "deep in the shamanistic past, and it is probable the *yoik* was not only a means to remember, but also a means to become. With the yoik a shaman might transpose his spirit into the shape of an animal or travel to far off places . . . the yoiker is opening himself to his subject, filling himself with it, in a sense he remembers by becoming."[29] This description is apt for Wajstedt, who seeks Sami rituals to immerse herself in traditions she missed and must recuperate, such as spending time up north during the reindeer branding, separation, and slaughter cycles, and also visiting large sacred Sami stones (known as *seitas*, signifying the nature gods to whom animal sacrifices were made). She feels ambivalent about filming this "holy" place—is it exploitative to do so, she asks? She compares the spooky feeling she gets in pursuit of it to the American film *The Blair Witch Project* and remarks that her car's windows seemed to open and close even with the car turned off. Near the conclusion of the film she ascends a steep slope to find the dark stones, which she calls an "entity," sees the fjord, and hears sounds of drums and rain, a northern encounter that must take place for the pilgrim as the final threshold. She takes us there too, but only so far, and then shuts off her camera. We understand the need for omission and silence, that yearning for what is most sacred and unknowable.

Conclusion: Survivance and Inspiration

For both filmmakers, their narratives—and the interviews and history embedded in these works—exemplify Vizenor's theory of "survivance." Rauna Kuokkanen calls survivance a proactive stance.[30] It is significant that the filmmakers present alternative ways of knowing and interpreting Sami experience from inside and outside, claiming objectification as a form of reflection, to create subjectivities in which they figure as storytellers with shamanic inklings. Going north and crossing borders are necessary transitions and entries to authenticity that may not be understood by outsiders and therefore are deemed threatening. Representing such experience through film enables connections, from the community, from intuition, and as an expression of inspired artistry. The result: a rich possibility for self-determination and a body of work for those who seek to understand where the Sami have come from and where they are headed. These films anticipate a powerful new time that belongs to the Sami, who also reach out now to other indigenous groups and make room for all of us.

We are presented with overlapping yet varied aspects of Sami history and subjectivity in these contemporary journeys through the self. Wajstedt and Kalvemo engage with public consciousness and evoke larger questions about political sovereignty and rights within and outside the northern Scandinavian and Russian Sami areas. Like other indigenous peoples, they embrace technology to examine the past and contemporary life and to offer new cultural formations that inspire, entertain, challenge, and enlighten. I look forward to continuing my journey deeper into the territory.

Acknowledgments

I wish to thank Peter Sköld and Krister Stoor for inviting me to the Rivers to Cross conference at the Center for Sami Research in Umeå, Sweden, in 2009, and for including an earlier version of this essay in the volume *Rivers to Cross: Sami Land Use and the Human Dimension*. (Umeå University, 2012). Also, I wish to thank the Department of Academic Affairs at Borough of Manhattan Community College, City University of New York, for a faculty development grant that enabled additional research in Norway.

Endnotes

1. In my analysis of Sami film, I am limited by my lack of knowledge of Sami and Scandinavian languages and by my outsider status, but I bring to bear a background in research and writing on ethnicity/gender/social class and environmental justice topics in film and literature. I've lived and worked in Scandinavia and spent time with Sami friends at conferences and festivals and in their homes and ancestral areas; I have visited the archives of Sami research centers and other libraries in Scandinavia. My focus has been on films by Sami directors that feature English subtitles. Since I began this research, Sami film has blossomed, and there are Sami film centers that support and promote the work, such as the International Sami Film Centre in Norway, and *Dellie maa*, a new Sami film festival that features film by indigenous peoples in northern Sweden.

2. Kathryn Burke, "The Sami Yoik," Sami Culture website, http://www.utexas.edu/courses/sami/diehtu/giella/music/yoiksunna.htm; Harald Gaski, "Introduction: Sami Culture in A New Era" and "Voice in the Margin: A Suitable Place for a Minority Literature," in *Sami Culture in a New Era: The Norwegian Sami Experience*, 9–28, 199–221 (Karsjok, Norway: Davvi Girji OS, 1997).

3. Liselotte Wajstedt, *Sami Nieida Jojk (Sami Daughter Yoik)* (Littlebig Productions in co-production with Filmpool Nord, Sirel Peensa, with support from Swedish Film Institute, 2007).

4. As I was revising this essay, Scandinavia House In New York City presented a film series, "Sami Side of Life: Films from the Far North" (June 4–18, 2014), with

selections from the Tromso International Film Festival, including *Sami Daughter Yoik*, as well as an art exhibit and lectures that ran through August 2014. In 2008, a program in Sami film was jointly presented by Scandinavia House and the National Museum of the American Indian. Liselotte Wajstedt has recently shown her films at the Imagine/Native Film + Media Arts Festival in Toronto and elsewhere.

5. Elina Helander and Kaarina Kailo, *No Beginning, No End: The Sami Speak Up*, The Canadian Circumpolar Institute Research Series No. 5 (published in cooperation with the Nordic Sami Institute, Finland, 1998).

6. Johs Kalvemo, *Frykten for De Grenselose* (*Fear of the Boundless*) (Norwegian Broadcasting Co./Sami Radio, 2002).

7. Gaski, *Sami Culture in a New Era*.

8. Victor Turner, "Liminality and Communitas," in *The Ritual Process: Structure and Anti-Structure* (Berlin: Walter de Gruyter, Inc., 1969), 95.

9. Rauna Kuokkanen, "Myths and Realities of Sami Women: A Post-Colonial Feminist Analysis for the Decolonization and Transformation of Sami Society," in *Making Space for Indigenous Feminism*, ed. Joyce Green (London: Zed Books, Fernwood Publishing, 2007).

10. Cheryl J. Fish, *Black and White Women's Travel Narratives: Antebellum Explorations* (Gainesville: University Press of Florida, 2004).

11. Lukas Allemann, *The Sami of the Kola Peninsula: About the Life of an Ethnic Minority in the Soviet Union* (University of Tromso Centre for Sami Studies, Skriftserie number 19, 2013).

12. Janike Kampevold Larsen and Peter Hemmersam, ch. 8 this volume.

13. Coppelie Cocq, *Revoicing Sami Narratives: North Sami Storytelling at the Turn of the 20th Century* (PhD diss. in Sami Studies, Umea University, 2008), 31–32.

14. Ibid.

15. Knut Hamsun, *Markenes Grode (Growth of the Soil)*, trans. by Sverre Lyngstad (New York: Penguin Books, 2007).

16. On the NBC TV sitcom *Welcome to Sweden* (2014), a Swedish character states that the Sami are a dilemma for his country, "our little Indians."

17. Helander and Kailo, *No Beginning, No End*, 10.

18. Wajstedt, *Sami Daughter Yoik*.

19. Gloria Anzaldúa, *Borderlands/La Frontera: The New Mestiza* (San Francisco: Spinsters/Aunt Lute, 1987).

20. Cocq, *Revoicing Sami Narratives*, 189.

21. Gaski, "Voice in the Margin," 202–203.

22. In 2008 Sami youth from four countries conducted a large peaceful protest in Rovaneimi, Finland, against the Finnish tourist industry for "using fake costumes and fake Samis" to greet foreign visitors at the Rovaneimi airport. They also objected

to the Finnish contestant in the Miss Universe pageant wearing a Sami *gakti*, which they decried as "showing a gross disrespect for Sami culture." Galdu Resource Center for the Rights of Indigenous Peoples.

23. Kalvemo, *Fear of the Boundless.*

24. Juhana Lumme, "Niillas Somby: A Wounded Healer" *Six Degrees* 3 (2007): 9.

25. Rauna Kuokkanen, "Indigenous Women in Traditional Economies: The Case of Sami Reindeer Herding," *Signs: Journal of Women in Culture and Society* 34, no. 3 (Spring 2009): 499–504.

26. Harald Eidheim, "Ethno-Political Development among the Sami after World War II," in Gaski, *Sami Culture in a New Era.*

27. Gaski, "Introduction" and "Voice in the Margin."

28. Wajstedt, *Sami Daughter Yoik.*

29. Hugh Beach, "Cultural Expression," in *Polar Peoples: Self-Determination and Development,* ed. the Minority Rights Group (London: Minority Rights Publications, 1994), 165.

30. Gerald Vizenor, quoted in Rauna Kuokkanen, "Survivance in Sami and First Nations Boarding School Narratives," *American Indian Quarterly* 27, no. 3–4 (Summer/Fall 2003): 699.

Bibliography

Anzaldúa, Gloria. *Borderlands/La Frontera: The New Mestiza.* San Francisco: Spinsters/Aunt Lute, 1987.

Allemann, Lukas. *The Sami of the Kola Peninsula: About the Life of an Ethnic Minority in the Soviet Union.* University of Tromso Centre for Sami Studies, Skriftserie number 19, 2013.

Beach, Hugh. "Cultural Expression." In *Polar Peoples: Self Determination and Development,* edited by the Minority Rights Group. London: Minority Rights Publications, 1994.

Burke, Kathryn. "The Sami Yoik." http://www.utexas.edu/courses/sami/diehtu/giella/music/yoiksunna.htm.

Cocq, Coppelie. *Revoicing Sami Narratives: North Sami Storytelling at the Turn of the 20th Century.* Doctoral dissertation in Sami Studies, Umea University, 2008.

Eidheim, Harald. "Ethno-Political Development among the Sami after World War II." In *Sami Culture in a New Era: The Norwegian Sami Experience,* edited by Harald Gaski. Karsjok, Norway: Davvi Girji OS, 1997.

Fish, Cheryl J. *Black and White Women's Travel Narratives: Antebellum Explorations.* Gainesville: University Press of Florida, 2004.

Gaski, Harald. "From Tundra to Desk." In *In the Shadow of the Midnight Sun: Contemporary Sami Prose and Poetry*. Seattle: University of Washington Press, 1998.

———. "Introduction: Sami Culture in A New Era" and "Voice in the Margin: A Suitable Place for a Minority Literature." In *Sami Culture in a New Era: The Norwegian Sami Experience.* Karsjok, Norway: Davvi Girji OS, 1997.

Hamsun, Knut. *Markenes Grode* (*Growth of the Soil*). Translated by Sverre Lyngstad. New York: Penguin Books, 2007.

Helander, Elina, and Kaarina Kailo. *No Beginning, No End: The Sami Speak Up*. The Canadian Circumpolar Institute Research Series No. 5. Published in cooperation with the Nordic Sami Institute, Finland, 1998.

Kalvemo, Johs. *Frykten for De Grenselose* (*Fear of the Boundless*). Norwegian Broadcasting Co./Sami Radio, 2002.

Kuokkanen, Rauna. "Survivance in Sami and First Nations Boarding School Narratives." *American Indian Quarterly* 27, no. 3–4 (Summer/Fall 2003).

———. "Myths and Realities of Sami Women: A Post-Colonial Feminist Analysis for the Decolonization and Transformation of Sami Society." In *Making Space for Indigenous Feminism*, edited by Joyce Green. London: Zed Books, Fernwood Publishing, 2007.

———. "Indigenous Women in Traditional Economies: The Case of Sami Reindeer Herding." *Signs: Journal of Women in Culture and Society* 34, no. 3 (Spring 2009): 499–504.

Lumme, Juhana. "Niillas Somby: A Wounded Healer." *Six Degrees* 3 (2007): 9.

Turner, Victor. "Liminality and Communitas." In *The Ritual Process: Structure and Anti-Structure*. Berlin: Walter de Gruyter, Inc., 1969.

Wajstedt, Liselotte. *Sami Nieida Jojk* (*Sami Daughter Yoik*). Littlebig Productions in co-production with Filmpool Nord, Sirel Peensa, with support from Swedish Film Institute, 2007.

12

Arctic Exposure

Nature, Race, and Regional Representation in Hollywood Film

susan kollin / MONTANA STATE UNIVERSITY

OF ALL THE OFFBEAT CHARACTERS WHO MAKE THEIR WAY TO Depression-era Hollywood in Nathanael West's 1939 novel *The Day of the Locust*, perhaps the most unexpected figures may be found in the Gingo family, the performing Eskimo trio hailing from Point Barrow, Alaska, who arrive in California to shoot retakes for a film about polar exploration. Although the picture had been released some time ago, the Eskimo father, mother, and son refuse to return to Alaska because, we are told, "[t]hey liked Hollywood."[1] The Gingo family manages to make a life for themselves in California, even finding the necessary substitutes for their arctic diet at Jewish delicatessens, where items such as smoked salmon, white fish, and pickled herring may be found. The cultural displacement experienced by newcomers from the Far North is alleviated here by the presence of California's long-established diasporic Jewish community, whose well-stocked shops enable the Gingos to create a new life outside their northern homeland.

The family's most noted appearance in the novel comes at a funeral for a clown, when they enter the chapel, bowing and waving to the grieving crowd as they make their way to their seats, like movie stars on opening night. The Gingos, we learn, are some of the only attendees who are actually eager to view the body of the deceased. "Only the Gingos stood up immediately," West writes. "They made for the coffin in a group . . . leaned over the coffin and told each other something in a series of thick, explosive gutturals. When they tried to take another look, Mrs. Johnson herded them firmly to their seats."[2] Like the endless stream of dreamers who come "to California to die," the Gingos are portrayed as being seduced by the illusion, artifice, and spectacle of fame associated with the movie industry.[3] Along with the starry-eyed but talentless ingenue Faye Greener, the washed-up screenwriter Tod Hackett, and the Native American salesman who hawks "genuine relics of the Old West" at "Tuttle's Trading Post," the seemingly misplaced Eskimo family represents the corruptions and shallowness of Hollywood's dream machine.[4] By portraying Eskimos in Hollywood as an absurd misplacement, the author deploys dominant modernist ideas about race, mobility, and geography to address a larger cultural mourning for the Real, an anxiety about what Rey Chow calls the "irreversibility of modernity," the profound metropolitan grief over rapid changes and the perceived loss of authenticity and nature in the twentieth century.[5]

A few years before Nathanael West's account of Eskimos in Hollywood was published, an Alaska Native actor made headlines in the Hollywood trade papers as an anomalous and intriguing new film celebrity. Iñupiaq actor Ray Mala, who may well have been West's model for the Gingo family, provides another instance in which an Eskimo came to Hollywood and decided to stay. From the 1930s to the 1950s, Mala's film career resulted in more than two dozen movies and television productions. The list includes lead roles in *Igloo* (1932), *Eskimo* (1934), and *The Last of the Pagans* (1935), as well as bit parts in *Union Pacific* (1939) with Barbara Stanwyck and Joel McCrea—released the same year that *The Day of the Locust* was published. It is unlikely that West as a writer of the Hollywood novel would have missed the media hype surrounding Mala, especially following the release of his second feature-length film, *Eskimo*, which garnered tremendous attention in 1934 with its Astor Theater opening in New York and was billed at the time as the "Biggest Movie Ever Made." Eventually dubbed the "Clark Gable of Alaska" in the trade papers, Mala emerged as a celebrity whose name would later become a chosen favorite among Iñupiaq families across arctic Alaska.[6] My argument here traces the ways dominant ideas about Alaska and its indigenous inhabitants in modern US culture have often relied on narrow understandings of nature, race, and place. Mala's Hollywood career, however, developed in unexpected ways that disrupt these cultural categories, and his experiences of travel, mobility, and cross-cultural affiliations offer new means of examining a denaturalized Alaska landscape as well as a more fluid sense of indigenous American identity for cultural critics.

Both West's treatment of the diasporic and cosmopolitan Gingo family in *The Day of the Locust* and the real-life fascination with Mala point to the ways modernist dislocation and mobility have often been understood in narrowly racialized ways. Examining what she calls "minority cosmopolitanism," critic Susan Koshy notes how an "unexamined logic" often shapes understandings of culture and mobility whereby the field of ethnic studies overlooks cosmopolitanism and diasporic studies in favor of arguments centered on the local and on national belonging.[7] She uses the term "minority cosmopolitanism" to address an experience of "diasporic citizenship" created within "a paradoxical relation to cross-cultural contact" that involves "the disruptions and asymmetries of intercultural encounter" as well as the "transformative possibilities" which emerge in these experiences.[8] In a similar way, in his study *Indians in Unexpected Places*, Philip Deloria examines long-held ideas about indigenous people and race, movement, and identity, particularly stereotypes about Native Americans who become naturalized in dominant discourse as timeless figures somehow consigned to a place of immobility. Focusing on the late nineteenth and early twentieth centuries, an era when indigenous peoples were nearly "dropped out of history," Deloria examines both the ways Native Americans engaged the same forces of modernization that also challenged non-Indians and the ways Indians were at the very center of rapid cultural changes even as they were endlessly pigeonholed as "primitive" peoples frozen in time and place.[9]

In this historical period, popular fascination with indigenous peoples who were thought to be unnaturally located "out of place" could be found in stories about the tragic fate of various Eskimos who were relocated to cosmopolitan centers across the United States and Europe. Shari Huhndorf notes, for instance, that in 1897 a public sensation arose concerning the experiences of six captive Eskimos held in New York as part of a museum display. After the surprising popularity of "live human exhibits" at the World's Columbian Exposition in Chicago in 1894, the anthropologist Franz Boas and other officials at the American Museum of Natural History asked the celebrated explorer Robert Peary to capture live human "specimens" for scientific study on his new voyage to the Far North. On the day after Peary's arrival in New York, twenty thousand visitors came to his ship, interested in catching a glimpse of the human captives on board. The individuals on display were not dressed for the high temperatures that day but instead were clad in the arctic fur that satisfied popular preconceptions of northern people. Later, these captive Eskimos would be "housed" in the basement of the American Museum of Natural History, where crowds of curious New Yorkers caught glimpses of them through a ceiling grate above their living spaces. Like nearly every instance of indigenous peoples forced to be on display throughout the history of Western imperial conquest, the Eskimos fell sick, and within a few months four of them died of pneumonia. As Huhndorf points out, "even in death, they endured prying Western eyes. The anthropologist Alfred Kroeber painstakingly documented the intimate display of grief and the death rituals of the survivors, as one by one, their companions and family members perished."[10] Such tragic stories of displacement and mobility often ended in death and contributed to popular ideas about the proper orderings of nature, race, and place.

Throughout histories of contact and encounter, Eskimos have frequently been used as a screen upon which Europeans and Euro-Americans project their own desires and anxieties about changing natures and cultures. As Fienup-Riordan notes, "Few people on earth have been written about so prodigiously or pictured so often in an exotic light" as Eskimos have been.[11] Dominant representations of Eskimos have typically centered on origins, she points out, particularly around the presumed lost origins of Western culture in the modern world. Understood as figures who embody social beginnings, Eskimos are often depicted as "peaceful, happy, childlike, noble, independent, and free."[12] In motion pictures especially, Eskimos have been figured as natural and "'essential man,' stripped of social constraints and High Culture."[13] In this way, they are meant to function in terms of what might be called the purest of "pure primitives," figures who are sentimentally elevated above the degrading compromises of a denaturalized modern era.[14] Drawing on this history of thinking,

West's portrait of the Eskimo family in *The Day of the Locust* as figures of an absurd displacement operates as a way of critiquing the seductive emptiness, delusional fantasies, and moral bankruptcy of modern culture as embodied especially in the corruption, shallowness, and artificiality of the Hollywood movie industry. By placing the Gingo family in this context, West points to what he sees as disturbing changes in the modern era that disrupt even the lives of the so-called purest of "pure primitives." As the author seems to imply, Alaska Natives like the Gingo family naturally belong in a place elsewhere, battling the harsh environmental elements of the Far North, not visiting film sets in the artificial, hypermodern world of sunny Hollywood.

Global Alaskans

In the case of Mala, what is interesting about his film experience is not merely his work as a cosmopolitan figure in Hollywood but also the variety of roles he played as an ethnic Other in a number of movies. Rather than limiting the parts he could play, Mala's racial identity for a time allowed him a certain degree of mobility in casting that likely prolonged his career in a world where film cycles moved in and out of popularity and where audiences grew quickly tired of what was once considered new. Eventually, he did face restrictions as the motion picture industry started adopting a code that forbade interracial love scenes, meaning that if he wanted to land a major romantic role it would need to be in a film with another "exotic."[15] In this way, although he was most famous in Hollywood in his roles playing Eskimos on screen, Mala was eventually cast broadly as a Hawaiian, a Tahitian, a Plains Indian, and other ethnic types in a variety of action-adventure films set in exotic, faraway lands. Such films highlighted US imperial desires for tales of adventure in wild, natural landscapes. Indeed, the cultural logic that allowed such casting and that enabled Mala to forge a Hollywood career emerged out of an American expansionist imaginary that saw exotic environments—and by extension, their indigenous inhabitants—as somehow interchangeable, a career that placed in him narratives located far and wide through what Amy Kaplan calls "the transnational routes of U.S. empire."[16]

Ray Mala was the first known Alaska Native to become a Hollywood movie star.[17] He was born in 1906 in the village of Candle, near Kotzebue, to an Iñupiaq mother and a Russian Jewish father who hailed from Siberia and was a prospector, trader, and hunter in Alaska. Orphaned at a young age after his father abandoned the family and his mother died in the great flu epidemic, Mala was educated at a Quaker missionary school in Kotzebue, where he learned English. He eventually ran away from the school and found work on board a number of expedition ships in the Arctic, including one with Danish explorer Knud Rasmussen. While on ship, he met a newsreel cameraman who took a liking to him. It was discovered that Mala was the only member on board the ship who could hand crank a camera in the extreme arctic temperatures, and he soon gained crucial experience as a cameraman. Later, the crewmember who had befriended him loaned Mala money to travel to Hollywood, where he found work in the movie industry.[18]

He was first known as Ray Wise when he began working as an assistant cameraman, then wrote a film script that he failed to sell, and finally became an actor in movies and television shows. Eventually he was selected by Universal Pictures to star in director Ewing Scott's *Igloo*, a film released in 1932 that was a remake of Robert Flaherty's *Nanook of the North*.[19] Later he was cast in W. S. Van Dyke's *Eskimo*, an adaptation of two books by famed arctic explorer Peter Freuchen. The actor took the name Mala, which was also the name of the character he played in the movie. In Hollywood he would have encountered a thriving group of Native American actors who were involved in the industry, primarily in the many Westerns that were being produced in the first half of the twentieth century. At this time, as Deloria explains, a number of Native American performers who had previously been involved in the Wild West shows had moved into the film industry, such that by the mid-twentieth century there existed a small but established community of American Indians living and working in Hollywood.[20] Early in his career, Mala's health suffered. He had frequent nose bleeds, complained of sore feet, experienced a bout of rheumatic fever, and also had high blood pressure. A doctor eventually recommended that he return to Alaska to recover, but Mala resisted the advice.[21]

Rather than being a "pure primitive" consigned to a fixed time and place, Mala was a cosmopolitan artist who traveled widely and witnessed events that vastly altered life in his homeland. Such experiences testify to how Alaska and the Arctic in general were far from being untouched blank spaces frozen in the past, just as these events also indicate how indigenous peoples were not somehow naturally fated to a life of stasis and immobility. Like his father who hailed from Russia, Mala experienced mobility within a world framed by imperial movements that began in the early eighteenth century with the first Russian attempts to colonize North America. Russian America eventually reached from present-day Alaska to as far south as Fort Ross, California.[22] In the eighteenth and nineteenth centuries, Alaska was the contested site of a number of European explorations in search of a Northwest Passage and as part of the fur trade and whaling industry, with the Spanish, French, British, and later Americans all making claims on the region's natural resources and leaving their marks on the map. In many ways, Mala's life was directly caught up in the history of the development of the Far North, from the travels of his Russian father across Alaska to his own encounters with two explorers who would greatly influence the direction of his future (Freuchen and Rasmussen), and finally to his journey to Southern California.

The Exotic North

Mala's film career also exposes a modern US fascination with exotic, faraway natures and peoples, beginning with the semiethnographic films *Igloo* and *Eskimo* and later with various South Sea adventure films he was cast in, including *The Last of the Pagans* (1935), an adaptation of Herman Melville's *Typee* that also featured Asian-American actress Lotus Long, *The Tuttles of Tahiti* (1942), and *Son of Fury: The Story of Benjamin Blake* (1942), and finally various Northwest melodramas, including *Call of the Yukon* (1938), *Girl from God's Country* (1940), and *Northwest Mounted Police* (1940). Such fascination with other landscapes and people was a common element of modern US identity. As Bill Brown has argued, during this period, a "wave of globalization in the U.S. (a result of immigration, middle-class travel, and international trade) provoked a romance of the local and an insatiable desire to know about innumerable locales. Most every region was produced *as knowledge* that could be incorporated, no matter how

Fig. 12.1. Ray Mala in the South Seas drama *Son of Fury: The Story of Benjamin Blake* (1942).

eccentric, into the national knowledge of itself—the eccentricity being an effect of the knowledge itself."[23]

Mala's interchangeability as a generic racial Other in Hollywood film was an offshoot of the modernist "romance of the local" and enabled the actor to compete for a variety of movie roles that extended his career for nearly two decades. One film in particular features a moment of unexpected rupture that challenges dominant US notions of race, nature, mobility, and place. The 1939 Cecil B. DeMille Western *Union Pacific* features Mala as a Sioux warrior (also named Mala) who is part of a raiding party that plunders a train making its way across Wyoming from Cheyenne to Laramie. The scene shows the main characters, played by Barbara Stanwyck and Joel McCrea, engaged in an unlikely and improbable rescue act whereby they

Fig. 12.2. Encounters with the cigar store Indian in *Union Pacific* (1939).

connect two ends of a broken telegraph line that has somehow made its way into the side of the train car to signal their trouble and request help back at the station.

In one scene, Mala picks through the contents of the derailed train and comes upon the figure of a cigar store Indian left in the wreckage. He first lifts up his tomahawk as if to protect himself against a potential enemy, his startled gaze at the strange carved wooden figure capturing his unsettled response at encountering the white man's Indian in this context (figs. 12.2 and 12.3). The history of the infamous cigar store Indian is worth noting here. The wooden figure was an invention of the advertising industry and emerged within a culture of illiteracy during a time when store owners relied on descriptive visual icons to popularize their merchandise. In

Fig. 12.3. Mirror poses in *Union Pacific* (1939).

the seventeenth century, European tobacco merchants used figures of American Indians to advertise their stores to a nonreading public. Most Europeans had never seen indigenous Americans, and many of the early cigar store Indians resembled enslaved Africans. By the time they arrived in North America in the early eighteenth century, the carved wood figures had taken on a more lifelike resemblance, although they were still based on dominant racial stereotypes that circulated about indigenous Americans in the era.[24]

In the scene, Mala uncovers a cigar store Indian that is meant to be transported by train to its new home as an advertising aid for a Wyoming tobacco merchant. While the character is at first startled by the figure, after a few moments he collects himself and calls his friends over to collect the wood image. The scene anticipates perhaps a line of thinking launched by Gerald Vizenor, who has aptly described what he calls the "kitschy" artifacts of Indianness in the history of American culture. Vizenor has famously argued that "the *indian* is an archive," a repository of colonial administration showing us nothing less than a false imperial history of "discoveries," "treaties," and "museum remains."[25] The scene features Mala the actor playing Mala the Plains Indian warrior and screens a moment of affinity between the Iñupiaq actor's appearance in the Western and the artificially constructed Indian figure he encounters in the wreckage alongside the train. In popular Westerns, the locomotive typically signals the arrival of modernity and thus the end of the West itself. In this way, Robin Murray and Joseph Heumann describe the appearance of railroads in their study *Gunfight at the Eco-Corral: Western Cinema and the Environment* as an "unbeatable opponent" of nature and an "unfettered vision of progress" in popular Westerns.[26] Here the train significantly goes off the track, its wreckage spewing forth representative elements of Anglo settler culture such as bloomers, a corset, a grandfather clock, various furniture, and the cigar store Indian, all of which fail to arrive at their final destination and thus do not serve their intended purposes in the history of progress—in this case, the history of Anglo-American nation building.

Vizenor might describe this cinematic moment of encounter as staging a post-Indian rupture that eschews the "romance of an aesthetic absence or victimry"[27] and that instead highlights "endurance and survivance," an insistence on a "sense of presence," perhaps even the construction of "a teasable self."[28] As Vizenor argues, "[d]oubtless, there are tricksters in the archives" that function as "the creases of an absence," which somehow manage to make themselves known in unpredictable and surprising ways.[29] In his recent book *The Inconvenient Indian: A Curious Account of Native People in North America,* critic and novelist Thomas King argues that Hollywood filmmakers tend to naturalize three basic Indian types—the bloodthirsty savage, the noble savage, and the dying savage.[30] In film and popular culture, Indians are known primarily through these racial stereotypes and clichés, or what King calls bits of "cultural debris," often embodied in objects such as war bonnets, beaded shirts, fringed deerskin dresses, loincloths, headbands, feathered lances, and so on.[31] For King, even with these basic

Indian types and clichéd elements of cultural debris, "[i]n the end, the history of Indians in Hollywood is more comedy than tragedy."[32] Such observations help make sense of this instance of racial recognition and affinity from *Union Pacific,* with its unexpected and strange humor disrupting what is otherwise not a very humorous film.

This fleeting and momentary example of rupture also challenges the genre of the popular Western, producing in turn what might be regarded as a "post-Western possibility," or an instance of contradiction and instability that opens up unexpected and new possibilities for the form. Neil Campbell describes post-Western films as texts that are both "haunted" and themselves haunt viewers by presenting them with various silences and absent presences.[33] Post-Westerns are "troubled and troubling texts about the West viewed no longer as an ideal, prelapsarian community . . . but rather as complex and awkward meanderings into a layered, scarred region."[34] As such, the post-Western cinematic moment may open up possibilities for "shifting beyond and engaging with the past, its discursive formations and its weighty frames."[35] We might extend these ideas by thinking of post-Westerns less as a genre and more as a sensibility. In the context of film noir, for instance, recent scholars such as Dan Flory suggest that we put aside conventional genre concepts and embrace instead a more fluid notion of "noir sensibility." By "sensibility," he refers to a "feeling or attitude" rather than a strict adherence to genre conventions in order to invoke the "features and effects" of the form.[36] Here Mala as both an actor and a character may help usher in a postindigenous and post-Western sensibility, reconfiguring ideas of nature, race, and mobility for modern movie audiences.

And perhaps the filmmakers were having a joke here at the Iñupiaq actor's expense—as if through his alleged inherent "natural" innocence, Mala would necessarily misrecognize the phony cigar store Indian as the real thing, someone who is related to him, in fact his "uncle," as he seems to be saying in the film. Yet the cinematic moment potentially carries other meanings, such that the joke may be read in reverse, providing instead an intervention into and reconfiguration of Hollywood's history of naturalizing racial stereotypes. Elizabeth Freeman discusses in another context what she calls the "time binds" of queer temporalities in ways that are useful to my discussion of this racial disruption in Hollywood cinema. In her study of the constructions of queer time and history, Freeman writes about the longing for "form" that often turns marginalized people "backward to prior moments, forward to embarrassing utopias, sideways to forms of being and belonging that seem, on the face of it, completely banal."[37] This movement involves a process whereby "as close enough readers of one another and of dominant culture," it becomes possible for marginalized populations "to gather up, literally, life's outtakes and waste products and bind them into fictitious but beautiful (w)holes."[38]

The scene certainly offers a new take on timely uses of the kitschy waste products of Anglo settler culture for marginalized people. In *Union Pacific*, Mala plays a character whose actions expose the underside of the genre and who escapes the imposed frames of nature, race, and genre in order to offer a postindigenous and post-Western performance. As such, Mala's multiple responses to the cigar store Indian, a figure that is itself a waste product of white settler culture and an instance of colonial debris in the film, marks a moment of critical possibility, an acknowledgment of the shared imperial histories of violence, misrepresentation, and marginalization that indigenous Americans have collectively faced throughout history.

Mala's career itself lays bare important insights into overlooked histories of forging new racial affinities, of dismantling constructed natures involving race and region, and of claiming cosmopolitan practices and identities for indigenous Americans across times and places. Ultimately, Mala's life and career also serve as an important corrective to popular US visions of the North as a place apart, an isolated retreat, or a blank geography. His work behind and in front of the camera provided him with opportunities to assert visibility and claim mobility, and thus with the critical means for breaking out of bounded or frozen stereotypes about northern lives, peoples, and places in ways that continue to have significance for audiences and scholars into the twenty-first century.

Endnotes

1. Nathanael West, *The Day of the Locust* (1939; New York: Signet, 1983), 122. The author had previously written "Three Eskimos," a sketch he did not publish but which served as the basis for his portrait of the Gingo family. The piece is collected in

Nathanael West: Novels and Other Writings, ed. Sacvan Bercovitch (New York: Library of America, 1997).

2. West, *Day of the Locust*, 124.

3. Ibid., 30.

4. Ibid., 164.

5. Rey Chow, *Writing Diaspora: Tactics of Intervention in Contemporary Cultural Studies* (Bloomington: Indiana University Press, 1993), 107.

6. Susan Hackley Johnson, "When Moviemakers Look North," *Alaska Journal* (Winter 1979): 17–18.

7. Susan Koshy, "Minority Cosmopolitanism," *PMLA* 126, no. 3 (May 2011): 592.

8. Ibid., 594.

9. Philip J. Deloria, *Indians in Unexpected Places* (Lawrence: University Press of Kansas, 2006), 6.

10. Shari Huhndorf, *Going Native: Indians in the American Cultural Imagination* (Ithaca: Cornell University Press, 2001), 79–80.

11. Ann Fienup-Riordan, *Freeze Frame: Alaska Eskimos in the Movies* (Seattle: University of Washington Press, 1995), xi.

12. Ibid.

13. Ibid.

14. Ibid., xii.

15. Lael Morgan, *Eskimo Star: From the Tundra to Tinseltown: The Ray Mala Story* (Kenmore, WA: Epicenter Press, 2011), 58.

16. Amy Kaplan, *The Anarchy of Empire in the Making of U.S. Culture* (Cambridge, MA: Harvard University Press, 2002), 61.

17. Evey Ruskin, "Memories of a Movie Star," *Alaska Journal* (Spring 1984): 34.

18. Morgan, *Eskimo Star*, 15 and 35; Ruskin, "Memories of a Movie Star," 34; and Johnson, "When Moviemakers Look North," 16.

19. Morgan, *Eskimo Star*, 43.

20. Deloria, *Indians in Unexpected Places*, 79–80.

21. Morgan, *Eskimo Star*, 46.

22. Ilya Vinkovetsky, *Russian America: An Overseas Colony of a Continental Empire, 1804–1867* (Oxford: Oxford University Press, 2014), 14, 46.

23. Bill Brown, *A Sense of Things: The Object Matter of American Literature* (Chicago: University of Chicago Press, 2004), 86.

24. "Cigar Store Indian," in *Encyclopedia of North American Indians*, ed. Frederick E. Hoxie (New York: Houghton Mifflin, 1996), 123.

25. Gerald Vizenor, *Fugitive Poses: Native American Indian Scenes of Absence and Presence* (Lincoln: University of Nebraska Press, 2000), 50.

26. Robin Murray and Joseph Heumann, *Gunfight at the Eco-Corral: Western Cinema and the Environment* (Norman: University of Oklahoma Press, 2012), 145.

27. Vizenor, *Fugitive Poses*, 15.

28. Ibid., 20.

29. Ibid., 114.

30. Thomas King, *The Inconvenient Indian: A Curious Account of Native People in North America* (Toronto: Doubleday Canada, 2012), 34.

31. Ibid., 54.

32. Ibid., 50.

33. Neil Campbell, *Post-Westerns: Cinema, Region, West* (Lincoln: University of Nebraska Press, 2013), 14.

34. Ibid., 15.

35. Ibid., 4.

36. Dan Flory, *Philosophy, Black Film, Film Noir* (University Park: Pennsylvania State University Press, 2008), 19–20.

37. Elizabeth Freeman, *Time Binds: Queer Temporalities, Queer Histories* (Durham: Duke University Press, 2010), xxii.

38. Ibid.

Bibliography

Brown, Bill. *A Sense of Things: The Object Matter of American Literature*. Chicago: University of Chicago Press, 2004.

Campbell, Neil. *Post-Westerns: Cinema, Region, West*. Lincoln: University of Nebraska Press, 2013.

Chow, Rey. *Writing Diaspora: Tactics of Intervention in Contemporary Cultural Studies*. Bloomington: Indiana University Press, 1993.

"Cigar Store Indian." *Encyclopedia of North American Indians*, edited by Frederick E. Hoxie. New York: Houghton Mifflin, 1996.

Deloria, Philip J. *Indians in Unexpected Places*. Lawrence: University Press of Kansas, 2006.

Fienup-Riordan, Ann. *Freeze Frame: Alaska Eskimos in the Movies*. Seattle: University of Washington Press, 1995.

Flory, Dan. *Philosophy, Black Film, Film Noir*. University Park: Pennsylvania State University Press, 2008.

Freeman, Elizabeth. *Time Binds: Queer Temporalities, Queer Histories*. Durham: Duke University Press, 2010.

Huhndorf, Shari. *Going Native: Indians in the American Cultural Imagination*. Ithaca: Cornell University Press, 2001.

Johnson, Susan Hackley. "When Moviemakers Look North." *Alaska Journal* (Winter 1979): 17–18.

Kaplan, Amy. *The Anarchy of Empire in the Making of U.S. Culture*. Cambridge: Harvard University Press, 2002.

King, Thomas. *The Inconvenient Indian: A Curious Account of Native People in North America*. Toronto: Doubleday Canada, 2012.

Koshy, Susan. "Minority Cosmopolitanism." *PMLA* 126, no. 3 (May 2011): 592–69.

Morgan, Lael. *Eskimo Star: From the Tundra to Tinseltown: The Ray Mala Story*. Kenmore, WA: Epicenter Press, 2011.

Murray, Robin, and Joseph Heumann. *Gunfight at the Eco-Corral: Western Cinema and the Environment*. Norman: University of Oklahoma Press, 2012.

Ruskin, Evey. "Memories of a Movie Star." *Alaska Journal* (Spring 1984): 34.

Vinkovetsky, Ilya. *Russian America: An Overseas Colony of a Continental Empire, 1804–1867*. Oxford: Oxford University Press, 2014.

Vizenor, Gerald. *Fugitive Poses: Native American Indian Scenes of Absence and Presence*. Lincoln: University of Nebraska Press, 2000.

West, Nathanael. *The Day of the Locust*. 1939; New York: Signet, 1983.

———. "Three Eskimos." *Nathanael West: Novels and Other Writings*, edited by Sacvan Bercovitch, 456–458. New York: Library of America, 1997.

13

Understanding Landscape Change Using Oral Histories and Tlingit Place-Names

daniel monteith / UNIVERSITY OF ALASKA SOUTHEAST

IMAGINE MASSIVE TIDAL GLACIERS CALVING OFF INTO THE OCEAN with large jets of water shooting up in the air and waves rolling at you while different types of shorebirds fly above and seals and whales swim around in the dark blue icy waters. This is an image many contemporary visitors to Alaska envision and seek out in Glacier Bay National Park. Numerous explorers and scientists fascinated by this rapidly changing, unusual landscape have studied the area and charted, mapped, and written about the bay and the several tidal glaciers now named Reid, Lamplugh, John Hopkins, Margerie, Grand Pacific, Muir, Riggs, and McBride. For the Tlingit who trace their ancestry to the bay, it represents a place of deep meaning and beauty. The Tlingit people have resided in and harvested resources from Glacier Bay "since time immemorial."[1] They have given their own names for each of the glaciers and hundreds of other places in the bay. The Tlingit maintain a close relationship to the land and possess personal names derived from the place-names. Tlingit speakers say Europeans name places after themselves but they name themselves after the places where they live and get their food. For the Tlingit people of Glacier Bay, it is foremost their homeland, their place of origin, their "breadbasket"[2] or sometimes called S'ix Tlein, meaning the "Big Dish."[3] One of the most common Tlingit names for the bay was Sit' eeti Geeyi, or "Bay in Place of the Glacier."[4] For most visitors Glacier Bay presents a place of scenic wonder and beauty. This essay offers a story of how researchers from anthropology, geology, biology, and other fields came together and collaborated with the Hoonah people, learning from Tlingit ecological knowledge, place-names, and oral histories to expand our knowledge and understanding of the human and natural history of Glacier Bay.

The dynamic glacial history of Glacier Bay provides a fascinating landscape to bring together different ways of knowing to tell a more holistic story than what you would hear on a tour of Glacier Bay National Park. Analyzing Tlingit ecological knowledge and oral histories alongside anthropology and geology tells a complete history, as Tlingit oral histories corroborate geological histories.

During the summers of 2003, 2004, and 2005, the University of Alaska Southeast (UAS) conducted fieldwork in the lower part of Glacier Bay. Assisting in the survey were National Park Service archaeologist Wayne Howell, retired NPS wildlife biologist Greg Streveler, and Ken Grant, NPS liaison to the community of Hoonah and an important Tlingit language specialist and advisor. As an anthropologist, I was a co-principal investigator with UAS geologist Cathy Connor. Richard and Nora Dauenhauer provided invaluable insights into the work and were important advisors. Throughout the rest of this chapter I will refer to this core group of researchers as the UAS team. While the UAS team anchored the geological work and ethnographic analysis, it is important to acknowledge other contributors, particularly the Tlingit ecological knowledge and historical work that rested squarely on the shoulders of the Hoonah elders and the community.

Geologists think of Glacier Bay as a dynamic landscape because of the presence of large ice fields and extensive glaciers. The advancing and retreating glaciers are constantly reconfiguring Glacier Bay. As the glaciers advance and retreat, they scour the landscape, leaving only geological clues about their movements. The glaciers' movements usually eliminate all archaeological clues about human activity. As glaciers advance and retreat, they can create a downward and upward lifting of the land in its path and around it. At the end of the Wisconsin ice age, the climate began to warm and glaciers retreated. As the glaciers melted, sea level rose, and as the weight of the glaciers receded up the valleys, the coastal lands began to rise. Geologists calculate that at the end of the Pleistocene, sea level rise may have been as much as three hundred to four hundred feet due to the melting of glacial mass along the coast. Near the end of the Pleistocene and beginning of the Holocene, ten thousand years ago, the glaciers in Glacier Bay rapidly retreated, transforming much of the bay area into saltwater fjords. In many respects, the bay would have looked as it does today.[5]

Approximately four thousand years ago, with cooling temperatures and the advance of the Little Ice Age (LIA), the network of glaciers slowly advanced and formed one large system. During the last several hundred years of the LIA, the glaciers rapidly advanced, until they pushed out to Icy Straits, arriving at their maximum point by the early 1700s. The clans of the Huna Kaawu have oral histories,[6] regalia, and songs that depict an event in which a glacier wipes out their village[7] near the end of the LIA. The nautical charts from 1794 and journals of Vancouver's voyage[8] of Southeast Alaska illustrate the initial and rapid retreat of the Grand Pacific Glacier.[9] John Muir's writings continue to document the constant and rapid retreat of the Grand Pacific Glacier throughout the late 1800s.[10] Today the Grand Pacific Glacier has retreated approximately sixty-five miles from its maximum terminus at the beginning of the 1700s. Recent geoarchaeological research indicates that the Huna oral history is at least four hundred years old.[11]

Archaeologists have found evidence of human activity and occupation just outside of Glacier Bay, in Groundhog Bay, that dates to just over ten thousand years ago. Over time, the Tlingit have observed and given place-names to many locations in the bay. The Tlingit used names that described the physical appearance of the landscape at a particular point in time. Many of these names have survived through oral histories for hundreds and possibly thousands of years. The Tlingit, in many instances, still use the older names rather than assigning new ones. The Tlingit names often describe the appearance of the ancient landscape rather than the modern-day reality. Thus, Tlingit place-names describe the landscape before the glaciers advanced, providing an opportunity for geologists to combine Tlingit knowledge with evidence left behind on the bedrock geology and in the soil sediments to interpret other details of the landscape.

Who Are the Tlingit?

The Tlingit tribe, or *k̲waan*, known as the Huna K̲aawu, is the Tlingit group that claims Glacier Bay as their ancestral homeland.[12] They still use the area and frequent the bay today. This area now makes up Glacier Bay National Park. The clans that make up the Huna people today are the Chookaneidi, Kaagwaantaan, Wooshkeetaan, and T'ak̲deintaan. Each clan is composed of several house groups. The Huna people have many oral histories about Glacier Bay.[13] The area in and around the modern-day national park represents land that is thought of as "places of origins" for the Huna Kaawu clans.

> For the people of Hoonah, Glacier Bay is At.oow—an owned place of abundant resources, clan origins, and territory. The Glacier Bay of former times is described by Hoonah Tlingits through their oral narratives, songs, place names, personal names, and clan and house crests. All of these things, including the land itself, are thought of in Tlingit culture as property which links the people to Glacier Bay historically, legally, and spiritually.[14]

At.oow is a concept often translated as meaning the property of a clan or house. Clan *at.oow* includes concepts like places, place-names, crests, clan regalia, *kooteeya* or carved poles, Tlingit personal names (that are handed down from generation to generation), songs, and the oral histories or narratives themselves.[15] This *at.oow* is brought out at potlatches or *ku.eexs* and is told and displayed to guests.[16] The oral histories have to be cited very precisely and specifically. This precision lends a great deal of reliability over many generations. In this practice Tlingit oral histories and narratives have been passed on for thousands of years. The songs, names, and regalia serve as memory devices that not only reinforce the memory but serve as a legal title to the clan's claims.[17]

Many Tlingit oral narratives address natural historical events, but one of the best known and most documented is the Sit' K̲aa Kax̲ Kana.aa, "Glacier Bay History." The Glacier Bay story has been recorded, transcribed, and translated by many visitors and scholars. The common element of the narrative is that it tells a story of how taboos were broken and the glacier was disrespected and then how a glacier wipes out a Tlingit village in Glacier Bay.

One of the first American visitors to record the Glacier Bay story was Eliza Scidmore,[18] who in the 1880s visited a small Tlingit village on Lester Island in Bartlett Cove, in the lower part of the bay. At the time she visited there was also a fish saltery at that village.[19] During the same period of Scidmore's visit to the Bartlett Cove Huna village, amateur ethnographer George T. Emmons was touring Southeast Alaska and recording many aspects of Tlingit culture and history. Emmons recorded some accounts of the Glacier Bay story and also took black-and-white photographs of the village. The UAS team used Emmons's photographs to examine how the landscape had changed due to historic activities and glacial rebound.[20] Following Emmons was a newly trained Boasian American anthropologist named John R. Swanton. Swanton recorded a few stories about Glacier Bay, including two stories about Kake'q!ute, which in a Tlingit relative chronology precede the events leading up to the glacier advancing to wipe out the village. He also recorded stories about the advance of the glacier.[21] One of the most detailed accounts of the Glacier Bay story is a version told by Susie James and recorded in Tlingit by the Alaska Native Language Center at the University of Alaska Fairbanks in 1973.[22] This version would be the inspiration for the Dauenhauers' transcriptions in the 1980s.[23] In her book *Do Glaciers Listen,* Julie Cruikshank includes different versions of the Glacier Bay story in her analysis of "local knowledge" and discusses Tlingit deference to glaciers.[24]

The Glacier Bay story contains details that can be very revealing to geologists. The narrative even describes the speed of the glacial advance: "It's growing like a dog running."[25] For geologists, the Sit' K̲aa Kax̲ Kana.aa story raises important questions: Does this story describe what is known as a surging glacier? When does the Sit' K̲aa Kax̲ Kana.aa narrative occur? What did the ancient landscape of Glacier Bay look like? Does the geological evidence align with the oral histories and Tlingit place-names? Is Glacier Bay a place where scientists can study landscape change using both geological science and Tlingit oral histories? These questions prompted the UAS team to use an interdisciplinary approach incorporating geoarchaeological, ethnohistorical, and ethnographic research. This approach corroborated both the science and the Tlingit oral histories and place-names.

Beginning in the early seventies, groundbreaking ethnographic and linguistic research began with consulting with Huna elders. In 1973, the

Sit' K̲aa Kax̲ Kana.aa Glacier Bay history was published by the Alaska Native Language Center at the University of Alaska Fairbanks. The narrator of the history was Kaasgeiy, or Susie James.[26] The narrative was recorded in Sitka in June 1972. Susie James was Chookan shaa of the Xinaa Hit (House Down the River). She was T'ak̲deintaan yadi, and her parents were Percy and Lilly Jackson. She was born on August 10, 1890, and died on November 3, 1980.[27] This research followed a general trend in anthropology and helped renew interest in using oral histories and biographies to understand the connection between culture and place.

The ethnographic work of the 1970s led to a major publication in the 1980s that recorded Tlingit narratives in both English and Tlingit, Nora and Richard Dauenhauer's *Haa Shuká, Our Ancestors: Tlingit Oral Narratives* (1987), which would become the first of their multiple-volume work on Tlingit narratives. In *Haa Shuká, Our Ancestors*, the Dauenhauers published two versions of the Glacier Bay history, one by Susie James and the other by Amy Marvin. The Marvin version was recorded on May 31, 1984, in Juneau. Amy Marvin was also Chookan shaa of the Naanaa Hit (Upper Inlet House) and T'ak̲deintaan yadi. Her mother's Tlingit name was Sxeinda.at. Marvin was a direct descendant of the two main figures in the story, Kaasteen and Shaawatseek.[28] These versions, which were published in both English and Tlingit, emphasized the importance of the native speakers' original narratives in the Tlingit language. The details of the Tlingit place-names and descriptions from these recorded narratives provide descriptive information about the landscape prior to the advance of the glaciers in Glacier Bay during the Little Ice Age.

Tlingit Place-Names and Ecological Knowledge

During the early 1990s, anthropologist Thomas Thornton began research on a Tlingit place-name study of Glacier Bay. In 1993, Thornton presented his research, titled "Tlingit and Euro-American Toponymies in Glacier Bay," at the Third Glacier Bay Science Symposium. This research was the beginning of a more comprehensive map project and traditional ecological knowledge (TEK) study. Throughout the decade, Huna elders elaborated and expanded on the Glacier Bay map and TEK study, and Huna elders and anthropological researchers continued to collect Tlingit place-names for the bay. By 2006, the Hoonah Indian Association (HIA), the federally recognized tribal council for the community of Hoonah, produced, copyrighted, and published a map, which is now for public sale.[29] The elders' contributions helped preserve the memory of the Tlingit place-names, and today educators use the names and map for language programs. Scholars and scientists of all disciplines can learn more about the natural history and how the landscape has changed through information derived from these names.

In 2003, I conducted qualitative ethnographic interviews of residents from Hoonah and Juneau with elders from the Huna Kaawu to provide more data about the oral histories, narratives, and use of Glacier Bay. Over the past several years many researchers have interviewed key consultants and elders about Glacier Bay. UAS students from Hoonah conducted interviews with their elders, providing an opportunity for them to learn more about their clan histories. When the interviewee gave permission, the interviews were recorded using both audiotape and videotape. The audiotapes, videotapes, and indexes of the interviews have been given to Hoonah Indian Association for archival purposes.

Tlingit place-names, which can be simultaneously historical and descriptive, provide a wealth of information about Glacier Bay. Some place-names are places that refer to the clan themselves, are their namesake, or are their "place of origin." Examples of this are the Chookaneidi, the people of Chookanheeni ("Grassy Place," Berg Bay, within Glacier Bay). Some place-names record historical events while others are names of clan members. Others refer to specific resources that can be found at the place. An example of this is Tleik̲w aani, "Berry Land or Place of Berries," which refers to a place in Glacier Bay where the Huna people went to pick berries.[30]

Place-names can also be descriptive terms. Many of these types of names describe the landscape, the appearance of the place in the past. L'eiwshaa Shake.aan refers to a place on top of the glacial sand dune. Today this area is a series of forested islands in the lower part of the bay; however, on these islands are dunes left over from the last glaciation. The name is still used even though the ancient descriptor does not depict the modern-day appearance.[31]

The details of the Tlingit place-names and descriptions from these recorded narratives may be crucial to a more holistic understanding of the landscape prior to the advance during the Little Ice Age. As the UAS project reconstructed the geoarchaeology of Glacier Bay, the local team carefully analyzed the Tlingit place-names and what the names suggested about the landscape. Some of the names gave us clues about different aspects of the landscape and terrain. Today, inside Glacier Bay, we can see plant succession in action. In the northern part of the bay are tidal glaciers and moraines. In the southern part of the bay, where we concentrated our research, glaciers had retreated far enough in the past that forest was developing. Many of the Tlingit names depict a landscape influenced by glaciers and reflect the fact that glaciers had been in much closer proximity. The following list of names reveals more the glacial impact and less the developing forest of today:

Tlingit place-name	**Translation/geographic explanation**
S'e Shuyee	Area at the end of the glacial silt
Sit'k'I T'ooch'	Little black glacier
Chookanheeni	Grassy River
Chookanheeni Yadi	Child of Chookanheeni, tributary stream
Ghatheeni	Sockeye salmon stream
Ghatheeni Tlein	Big Sockeye Salmon River
Aax'w Xoo	Among the Little Lakes
L'eiwshaayi	Sand mountain
L'eiwshaa Shake Aan	Town on top of the glacial sand dunes
T'ooch Ghi'l'i	Black cliff
Tleiw shayee	Clay point

Tlingit Oral Histories and Deep History

The oral histories and place-names of the Tlingit provide a diachronic perspective to their occupation in Southeast Alaska. Many scholars have attempted to put a Euro-American chronology to the antiquity of Tlingit occupation in the region. Some research would position the Tlingit and their early oral histories in the context of the end of the Pleistocene, beginning of the Holocene, or ten thousand years ago. While some conservative archaeologists might question this depth of antiquity, most today agree that the Tlingit have lived in Southeast Alaska for hundreds, if not thousands, of years. Within a Tlingit relative chronology, the Glacier Bay story is a recent one. Some Tlingit narratives most likely predate the story of the glacial advance. These earlier oral histories also may give us more clues about the ancient landscape. Most of the Tlingit clans have oral histories that were labeled "flood" stories by early missionaries and schoolteachers. When our research group examined these histories, we found that the narratives may be describing "deep history,"[32] or events that occurred at the end of the Pleistocene and beginning of the Holocene, ten thousand years ago.

Tlingit oral narratives, like many indigenous narratives, are a combination of different genres, including oral histories, dream time stories, visions, and epic creation stories, just to name a few. Many Tlingit narratives could also be described as metaphors for explaining natural historical events and processes. Simply put, some of these Tlingit narratives have specific historical events interwoven with other elements.

Some of the challenges to using and understanding Tlingit oral histories are similar to using any oral historical traditions to complement other sources. Each generation of teller within Tlingit culture is coached to tell the story as accurately and as precisely as possible, but each storyteller may ever so slightly add their nuance or perspective to the story. Thus, each clan may tell the story slightly differently, reflecting their own version based on their own unique history or the clan's perspective during the event. Over the years the Glacier Bay history has been recorded in different languages with different translators or transcribers. How each person recorded the story no doubt had an impact. Did the person recording the story write it down or use a recording device, which could have influenced the recording and the storyteller? When examining many versions of the Glacier Bay story, some of the recorders condensed a number of details, and their own biases or knowledge influenced the condensation of the story. Also, the interpretation by listeners and/or readers plays a role in what is understood about the story. Understanding Tlingit oral histories both for the present-day listener and for the storyteller may be a little confusing when the physical appearance of the landscape is changing rapidly.

Interdisciplinary research can enhance understanding of both how the landscape is changing as well as the tensions of maintaining the stories and making them applicable to the present environment. The geoarchaeological component of the UAS team research concentrated on the lower part of the bay, for two reasons. First, it was the most likely location for the village known as Ghatheeni, which was wiped out by the glacier in the Glacier Bay story. Second, the upper part of the bay had already received a great deal more research attention because of the presence of the glaciers. Much of the previous research focus had been in Muir Inlet and the upper regions of the bay by the Institute of Polar Studies at Ohio State University. In the mid-1960s, Goldthwaite[33] and Haselton[34] focused their research in Muir Inlet. These researchers collected soil samples and wood from intertidal tree stumps and began accumulating radiocarbon dates that enabled them to develop a chronology for the advance of the glaciers in Muir Inlet. Researchers from the Institute of Polar Studies continued this fieldwork, with Mickelson working around the Burroughs Glacier[35] and McKenzie and Goldthwaite concentrating on Adams Inlet.[36] In the 1980s, Goodwin worked in Muir Inlet, where his research collected "glaciolacustrine" sediments and provided finer resolution of the glacial advance during the Holocene.[37] Because this previous research was focused closer to the contemporary presence of the glaciers in the upper bay, the UAS team was able to build off this research and radiocarbon dates. We used the dates collected by these earlier research projects to construct a GIS database and computer-animated maps. We then made different maps from specific periods, creating snapshots in time of the glacial advance. The UAS team decided to pursue its fieldwork in the lower part of the bay to continue the study of the advance of the glaciers during the Little Ice Age.

The UAS team's work contributed to the understanding of the lower part of the bay and potentially to the development of a chronology for the advancement of the glacier during the Neoglacial or Little Ice Age. In the lower part of the bay there are many islands, which the geoarchaeological research team explored via kayaks. We were looking for intertidal stumps from trees that were killed or possibly run over by the advancing glacier. A small sample of wood was collected from a few select tree stumps, and then these samples were radiocarbon dated. Each of the dates gave us data about when the glacier had advanced to each specific area. Cut banks and stream banks were explored and analyzed for soil profiles. We collected flora and faunal samples, which provided both information about the landscape and organics that could be radiocarbon dated. We collected a total of ten samples that we had radiocarbon dated. The following is a table of the places, dates, and types of material dated:

Location	Date (Carbon 14 Age)	Material dated	Beta #
Kidney Island	4310 +/– 40 BP	Shell	194096
Head of Berg Bay	4380 +/– 50 BP	Shell	194100
S. Shore Berg Bay	2300 +/– 40 BP	Wood	194103
S. Shore Berg Bay	2120 +/– 40 BP	Wood	194102
Rush Point	1860 +/– 40 BP	Wood	194104
Head of Berg Bay	1910 +/– 60 BP	Wood (peat)	194101
Fox Farm Island #3	1630 +/– 60 BP	Wood	194099
Kidney Island	1300 +/– 50 BP	Wood	194098
Kidney Island	430 +/– 60 BP	Wood	194097
Lester Island	370 +/– 50 BP	Wood	194095

The geological samples from different locations in the bay gave us absolute dates and information about the landscape. The samples are listed by location from north to south, the same direction of the advance of the glacier. As one can see, the dates descend in age from north to south at a gradual chronological progression.

With the radiocarbon dates from the geoarchaeological samples and GIS modeling, we were able to re-create the advance of the glacier. Moreover, inferring from the types of floral and faunal samples collected we were able to develop a conceptual model about the appearance of the landscape (that is, terrestrial or marine) over time. One big question concerned the ecology or appearance of the landscape at the terminus, or foot, of the glacier. In other words, did the stage setting for the Kaasteen story corroborate with the geoarchaeological samples? The materials collected in the lower part of the bay were terrestrial deposits, not marine deposits. In the final analysis, the samples confirmed a terrestrial scene as the glacier advanced to the village of Ghatheeni. Tlingit stories refer to the village sometimes as Ghatheeni or adjacent to Ghatheeni, which translates to "sockeye stream." We were able to surmise that there was a sockeye creek, Ghatheeni, both before and after the advance of the glacier. We were not able to establish the exact location of Ghatheeni in the ancient landscape. But the mere discussion of the presence both before and after the glacial advance provoked discussion and research among fishery biologists about the exodus and recolonization of sockeye on different streams in the bay. GIS modeling of the ancient landscape may provide some clues about some of the DNA information fish biologists are attempting to identify. Understanding how streams change and why salmon recolonize different streams has applications today. Many of the glaciers in Southeast Alaska are receding. Receding glaciers impact different characteristics of the land and streams, including uplift of the land, stream temperatures, stream turbidity, water flow, distance, and gradient to saltwater.

Conclusions

From the lessons learned in Glacier Bay, we have integrated geology, archaeology, GIS modeling, Tlingit place-names, and oral histories to better understand landscape change elsewhere in Southeast Alaska. The research being conducted in other parts of the region has enabled scientists to develop better predictive models that may help us find older cultural sites in Southeast Alaska. Archaeological or cultural sites more than six thousand years old have been very hard to find in the region, primarily due to the significant and variable rate of glacial or isostatic rebound. Many of the older sites are a significant distance and elevation from present-day shorelines or beaches. Another application of this interdisciplinary research is using these techniques to better understand landscape change, soil formation, and forest productivity.

In a region where the land is uplifting at a rapid pace, further research utilizing Tlingit place-names and oral histories may enable scientists to develop better predictive models for finding early Holocene sites.[38] Many oral histories tell about earthquakes, tsunamis, volcanic eruptions, glacial events, and floods. Another productive avenue of research could utilize Tlingit ecological knowledge to help scientists understand more about changes in seasonality and abundance and availability of flora and fauna due to climate change and economic development. This interdisciplinary research has been rewarding because it has established credibility of Tlingit oral histories and traditional ecological knowledge, demonstrating how different ways of knowing can enrich our understanding of landscape change and cultural resilience.

While Tlingit science and knowledge have benefited Euro-American science, "Western" science is also helping to advance indigenous knowledge and research. This recent interdisciplinary research has been integrated into classrooms for all levels in Southeast Alaska and lends credence to these Tlingit oral histories. The Glacier Bay Tlingit map has provided teaching material for language learners of all levels. At least three of the Tlingit students who participated in this research are now teachers in the public schools.

The Tlingit place-name map of Glacier Bay has also worked as a political tool to assist the people of Hoonah and Yakutat to begin to plant the seeds to decolonialize Euro-American names from Glacier Bay National Park. Within the park, several sites are potential traditional cultural properties, based on the guidelines established in the National Historic Preservation Act. These places, like Chookanheeni, are thought to be places of origin and are namesakes for the clan. The Tlingit place-names and oral histories provide important documentation for these sites in the nomination process, as they possess little or no tangible archaeological or physical evidence because it was erased by glacial activity. Thus it is Tlingit oral histories and place-names that leave the indelible voices of the ancestors on the land. The fruits of this collaborative interdisciplinary research promoted new levels of awareness and ways of thinking that led to a deeper understanding of the human and natural history of Southeast Alaska. We need to be taking these histories more seriously if we hope to protect the ecological integrity of the North, which is often framed as the "last" frontier or "last" bit of wilderness on the planet. Caring about that wilderness is futile without these ways of knowing, and the languages that carry them.

Endnotes

1. Daniel Monteith, "Tongass, the Prolific Name, the Forgotten Tribe: An Ethnohistory of the Taantakwaan Tongass People" (PhD diss., Department of Anthropology, Michigan State University, 1998), 53.

2. Walter R. Goldschmidt and Theodore H. Haas, *Haa Aani, Our Land: Tlingit and Haida Land Rights Use,* ed. with an introduction by Thomas F. Thornton (Seattle: University of Washington Press, 1998), 54, 131.

3. Thomas F. Thornton, "Tleiḵw Aani, the 'Berried' Landscape: The Structure of Tlingit Edible Fruit Resources at Glacier Bay, Alaska," *Journal of Ethnobiology* 19 (1999): 29.

4. Thomas F. Thornton, "Tlingit and Euro-American Toponymies in Glacier Bay," *Proceeding of the Third Glacier Bay Science Symposium, 1993* (Anchorage: US Department of the Interior, National Park Service, 1993), 294.

5. Cathy Connor, Greg Streveler, Austin Post, Daniel Monteith, and Wayne Howell, "The Neoglacial Landscape and Human History of Glacier Bay, Glacier Bay Park and Preserve, Southeast Alaska, USA," *Holocene* 19, no. 3 (2009): 375–387.

6. Susie James (Kaasgeiy X̱'eidax̱), *Sit'Ḵaa Kax̱ Kana.aa (Glacier Bay History)*, ed. and transcribed by Nora Florendo (Sitka: Tlingit Readers, in association with the Alaska Native Language Center, 1973), 1–23.

7. Nora Marks Dauenhauer and Richard Dauenhauer, eds., *Haa Shuká, Our Ancestors: Tlingit Oral Narratives* (Seattle: University of Washington Press, 1987).

8. George Vancouver, *A Voyage of Discovery to the North Pacific Ocean and Round the World, performed in the Years of 1790–1795*, ed. and annotated by W. Kaye Lamb (London: The Hakluyt Society, 1984), map appendices.

9. Archibald Menzies, *The Alaska Travel Journal of Archibald Menzies 1793–1794*, ed. Wallace M. Olson (Fairbanks: University of Alaska Press, 1993).

10. John Muir, *John Muir Travels in Alaska* (Boston: Houghton Mifflin Company, 1979).

11. Connor et al., "Neoglacial Landscape and Human History," 375–387.

12. Walter R. Goldschmidt and Theodore H. Haas, *Possessory Rights of the Natives of Southeastern Alaska: A Report to the Commissioner of Indian Affairs* (Washington, DC, 1946).

13. John R. Swanton, "Social Condition, Beliefs, and Linguistic Relationship of the Tlingit Indians," in *Bureau of American Ethnology Twenty-sixth annual report*, 393–512 (Washington, DC, 1908).

14. Daniel Monteith, Cathy Connor, Wayne Howell, and Greg Streveler, "Geology and Oral History-Complimentary Views of a Former Glacier Bay Landscape," in *Proceedings of the Fourth Glacier Bay Science Symposium*, ed. John F. Piatt and Scott M. Gende (Anchorage: National Park Service, 2007), 50.

15. Nora Dauenhauer, "Tlingit At.oow: Traditions and Concepts," in *The Spirit Within*, ed. Steve Brown (Seattle: Seattle Art Museum, 1995), 15.

16. Sergei Kan, *Symbolic Immortality* (Washington, DC: Smithsonian Institution Press, 1989), 43–46.

17. Daniel Monteith, "Tlingit Oral Narratives and Time Immemorial," paper presented at the Conference of the Tlingit and Haida Clans, March 22–25, 2007, Sitka, Alaska, p. 2.

18. Eliza R. Scidmore, *Journeys in Alaska* (Boston: D. Lothrop and Company, 1885).

19. Eliza R. Scidmore, "Glacier Bay," in *Appleton's Guide to Alaska and the Northwest Coast* (New York: D. Appleton and Company, 1899).

20. George Emmons, *The Tlingit Indians*, ed. Frederica de Laguna (Seattle: University of Washington Press, 1991), 58, 368.

21. John R. Swanton, *Tlingit Myths and Texts*, Bureau of American Ethnology Bulletin 39 (Washington, DC: US Government Printing Office, 1909).

22. James, *Sit'K̲aa Kax̲ Kana.aa*, 1–23.

23. Dauenhauer and Dauenhauer, *Haa Shuká, Our Ancestors.*

24. Julie Cruikshank, *Do Glaciers Listen? Local Knowledge, Colonial Encounters, and Social Imagination* (Vancouver: University of British Columbia Press, 2005).

25. Dauenhauer and Dauenhauer, *Haa Shuká, Our Ancestors*, 253.

26. James, *Sit'K̲aa Kax̲ Kana.aa.* 1–23.

27. Dauenhauer and Dauenhauer, *Haa Shuká, Our Ancestors.*

28. Ibid.

29. Hoonah Indian Association, *Tlingit Place Names of the Huna Kaawu* (Hoonah, AK: Hoonah Indian Association, 2006).

30. Thornton, "Tleik̲w Aani, the 'Berried' Landscape," 27.

31. Connor et al., "Neoglacial Landscape and Human History " 375–387.

32. Monteith, "Tlingit Oral Narratives," 2.

33. R. P. Goldthwaite, F. Loewe, F. C. Ugolini, H. F. Decker, D. W. DeLong, M. R. Trautman, E. F. Good, T.R.I Mereli, and E. D. Rudolph, *Soil Development and Ecological Succession in a Deglaciated Area of Muir Inlet, Southeast Alaska*, Institute of Polar Studies Report no. 20 (Columbus: Ohio State University, 1966).

34. G. M. Haselton, *Glacial Geology of Muir Inlet, Southeast Alaska*, Institute of Polar Studies Report no. 18 (Columbus: Ohio State University, 1966).

35. D. M. Mickelson, *Glacial Geology of the Burroughs Glacier Area, Southeastern Alaska*, Institute of Polar Studies Report no. 40 (Columbus: Ohio State University, 1971).

36. G. D. McKenzie and Goldthwaite, R. P., "Glacial History of the Last Eleven Thousand Years in Adams Inlet, Southeastern Alaska," *Geological Society of America Bulletin* 82 (1971): 1767–1782.

37. R. G. Goodwin, "Holocene Glaciolacustrine Sedimentation in Muir Inlet and Ice Advance in Glacier Bay, Alaska, USA," *Arctic and Alpine Research* 20 (1988): 55–69.

38. Daniel Monteith, "Current Transdisciplinary Research on the Natural History and Landscape Change in Southeast Alaska," in *Proceedings from the Alaska Historical Society (AHS) Annual Meeting, September 26–28, 2013* (Haines: Alaska Historical Society 2013).

Bibliography

Connor, Cathy, Greg Streveler, Austin Post, Daniel Monteith, and Wayne Howell. "The Neoglacial Landscape and Human History of Glacier Bay, Glacier Bay Park and Preserve, Southeast Alaska, USA." *Holocene* 19, no. 3 (2009): 375–387.

Cruikshank, Julie. *Do Glaciers Listen? Local Knowledge, Colonial Encounters, and Social Imagination*. Vancouver: University of British Columbia Press, 2005.

Dauenhauer, Nora. "Tlingit At.oow: Traditions and Concepts." In *The Spirit Within*, edited by Steve Brown. Seattle: Seattle Art Museum, 1995.

Dauenhauer, Nora Marks, and Richard Dauenhauer, eds. *Haa Shuká, Our Ancestors: Tlingit Oral Narratives*. Seattle: University of Washington Press, 1987.

Emmons, George. *The Tlingit Indians*. Edited by Frederica de Laguna. Seattle: University of Washington Press, 1991.

Goldschmidt, Walter R., and Theodore H. Haas. *Haa Aani, Our Land: Tlingit and Haida Land Rights Use*. Edited with an introduction by Thomas F. Thornton. Seattle: University of Washington Press, 1998.

———. *Possessory Rights of the Natives of Southeastern Alaska: A Report to the Commissioner of Indian Affairs*. Washington, DC, 1946.

Goldthwaite, R. P., F. Loewe, F. C. Ugolini, H. F. Decker, D. W. DeLong, M. R. Trautman, E. F. Good, T.R.I. Mereli, and E. D. Rudolph. *Soil Development and Ecological Succession in a Deglaciated Area of Muir Inlet, Southeast Alaska*. Institute of Polar Studies Report no. 20. Columbus: Ohio State University, 1966.

Goodwin, R. G. "Holocene Glaciolacustrine Sedimentation in Muir Inlet and Ice Advance in Glacier Bay, Alaska, USA." *Arctic and Alpine Research* 20 (1988): 55–69.

Haselton, G. M. *Glacial Geology of Muir Inlet, Southeast Alaska*. Institute of Polar Studies Report no. 18. Columbus: Ohio State University, 1966.

Hoonah Indian Association. *Tlingit Place Names of the Huna Kaawu*. Hoonah, AK: Hoonah Indian Association, 2006.

James, Susie (Kaasgeiy X̲'eidax̲). *Sit'K̲aa Kax̲ Kana.aa (Glacier Bay History)*. Sitka: Tlingit Readers in association with the Alaska Native Language Center, 1973.

Kan, Sergei. *Symbolic Immortality*. Washington, DC: Smithsonian Institution Press, 1989.

McKenzie, G. D., and Goldthwaite, R. P. "Glacial History of the last Eleven Thousand Years in Adams Inlet, Southeastern Alaska." *Geological Society of America Bulletin* 82 (1971): 1767–1782.

Menzies, Archibald. *The Alaska Travel Journal of Archibald Menzies 1793–1794*. Edited by Wallace M. Olson. Fairbanks: University of Alaska Press, 1993.

Mickelson, D. M. *Glacial Geology of the Burroughs Glacier Area, Southeastern Alaska*. Institute of Polar Studies Report no. 40. Columbus: Ohio State University, 1971.

Monteith, Daniel. "Current Transdisciplinary Research on the Natural History and Landscape Change in Southeast Alaska." In *Proceedings from the Alaska Historical Society (AHS) Annual Meetings, September 26–28, 2013*. Haines: Alaska Historical Society, 2013.

———. "Tlingit Oral Narratives and Time Immemorial." Paper presented at the Conference of the Tlingit and Haida Clans, March 22–25, 2007, Sitka, Alaska.

———. "Tongass, the Prolific Name, the Forgotten Tribe: An Ethnohistory of the Taantakwaan Tongass People." PhD diss., Department of Anthropology, Michigan State University, 1998.

Monteith, Daniel, Cathy Connor, Wayne Howell, and Greg Streveler. "Geology and Oral History-Complimentary Views of a Former Glacier Bay Landscape." In *Proceedings of the Fourth Glacier Bay Science Symposium*. Edited by John F. Piatt and Scott M. Gende. Anchorage: National Park Service, 2007.

Muir, John. *John Muir Travels in Alaska*. Boston: Houghton Mifflin Company, 1979.

Scidmore, Eliza R. *Journeys in Alaska*. Boston: D. Lothrop and Company, 1885.

———. "Glacier Bay." In *Appleton's Guide to Alaska and the Northwest Coast*. New York: D. Appleton and Company, 1899.

Swanton, John R. *Tlingit Myths and Texts*. Bureau of American Ethnology Bulletin 39. Washington, DC: US Government Printing Office, 1909.

———. "Social Condition, Beliefs, and Linguistic Relationship of the Tlingit Indians." In *Bureau of American Ethnology Twenty-sixth annual report*, 393–512. Washington, DC, 1908.

Thornton, Thomas F. "Tleik̲w Aani, the 'Berried' Landscape: The Structure of Tlingit Edible Fruit Resources at Glacier Bay, Alaska." *Journal of Ethnobiology* 19 (1999).

———. "Tlingit and Euro-American Toponymies in Glacier Bay." In *Proceedings of the Third Glacier Bay Science Symposium*. Anchorage: United States Department of the Interior, National Park Service, 1993.

Vancouver, George. 1984. *A Voyage of Discovery to the North Pacific Ocean and Round the World, performed in the Years of 1790–1795*. 4 vols. with introduction and appendices. Edited and annotated by W. Kaye Lamb. London: The Hakluyt Society, 1984.

14

Prospecting for Buried Narratives in Wrangell-St. Elias National Park and Preserve

margot higgins / MACALESTER COLLEGE

What shall we do with our histories?

We must confess our histories. We must acknowledge our histories. We must remember our histories. It is only after those things are done that we can begin to heal, and to communicate with one another as the equals that we are.

Let us tell each other stories that underscore the parallel nature of our lives. Let us tell each other stories that value the evident relationship in our condition. Let recognition and respect and knowledge travel in both directions.

—*Ernestine Hayes, Tlingit Elder and Professor of English, University of Alaska Southeast*

IN ALASKA'S WRANGELL–ST. ELIAS NATIONAL PARK AND PRESERVE, through narratives told on the Internet, in tourist brochures, on roadside interpretive signs, and by park staff, the National Park Service (NPS) tells a familiar boom-and-bust narrative that is focused on the area's brief industrial mining era from 1900 to 1938. One of those larger narratives relayed by the NPS and a number of local park residents celebrates the railway, developed by the Kennecott Corporation to take some of the richest copper ore in the world from its origins in the Wrangell Mountains to the coast in Cordova. Built by New York capital investors and thousands of workers, who laid 196 miles of tracks around glaciers, across canyons, and through deep snow and avalanche areas, the railroad carried two hundred million tons of copper ore from 1911 to 1938, when the Kennecott mines closed. It was copper from Kennecott, carried by these corporate trains on this treacherous route, that charged the nation with a new form of electricity at the beginning of the twentieth century.[1]

In the summer of 2011, the railway completion story was conveyed during a weekend-long centennial celebration that included a historical slide show, a reenactment of the railroad opening, a period dress party, and tea and cake on the lush lawn of an upscale local lodge. Just a few days before the festivities, the local historical museum and the NPS added to the agenda a dance recital to be performed by local Ahtna children, most of whom are now living in low-income residential communities outside park boundaries. At the conclusion of a performance that held the attention of a packed lecture hall of park residents and tourists, the interpretive supervisor for Wrangell–St. Elias bestowed a gift upon the Ahtna people: a copper railroad spike. His intentions very well may have been a sincere effort to demonstrate a gesture of appreciation for the time, travel, and years of practice involved in the performance, but such an offering might also be interpreted as another sharp nail in a long-concealed coffin of cultural erasure.

To an audience well accustomed to a dominant cultural narrative that reinforces Western language, perceptions, and values, such an act did not raise any substantial critique. In order to understand the cultural insensitivity that was displayed in Kennecott on the day of the centennial, one must dig beyond railroads and tea parties into the economic and cultural histories of this area. One needs to understand the making of this landscape, how it has been shaped by race, how it continues to be influenced by past and present colonial approaches and economic transition. Examining the interaction of local and global narratives about human and ecological history from various park residents and managers can reveal more nuanced social, historical, political, and economic factors that affect national park management but have so far been more elusive to NPS managers.

In this chapter I will consider the cultural specificity of the Kennecott mining narrative, examine how particular histories have been left out of mainstream narratives of the area, and consider some of the larger consequences of that erasure. Several scholars have also examined how an ongoing frontier mentality applies to national park management and how indigenous voices have been left out.[2] Yet they have not adequately examined how NPS narratives that reinforce national myths and identities for the sake of tourism combine with local narratives and responses. Resynthesizing the interaction between these narrative scales may allow for a more thorough interpretation of the connections among national parks, past and ongoing human displacement, and uneven economic gain. What follows is a narrative account of the complex material and discursive relationships between NPS, local peoples of Euro-American descent, indigenous Alaskans, and the construction of the area's mining history, human history, and natural history.

Theoretical Framing

Critical theorist Bruce Willems-Braun has argued that streams of past colonialism continue to infuse the present.[3] He developed the influential idea of "buried epistemologies," or the conception that certain patterns of thinking are inherited and often contribute to unacknowledged discrimination and social inequity. Euro-American theories of knowledge, in particular, have often inherited particularly exclusive colonial ways of making sense of the world. As Jake Kosek[4] argues, moreover, people remake the past through their memories and they bring these meanings to bear on present conditions. To compare these past and present perspectives, Julie Cruikshank contends that narratives, particularly in the form of oral histories, "provide an observatory from which to assess the shifting boundary between what we call *history* and what we call *myth*."[5] Carolyn Finney builds on Cruikshank's work, claiming that narratives "inform our environmental interactions and shape the institutions concerned with environmental issues."[6] As a key manager and arbiter of national environmental attitudes, the National Park Service is no exception in crafting its narratives. To be sure, as the critical parks literature details, the history of the ideals of preservation and nationalism that surround the establishment of parks are simultaneous with and contingent upon the displacement of both white settlers on indigenous lands and the previous colonial intrusions that made these dispossessions possible.[7]

Thus, narratives might provide insight into intentional or nonintentional decisions made by the NPS, as one institution through which we can view the continued presence of colonialism and the ways that it continues to shape national perceptions that reinforce discrimination and inequality. Narratives also provide a critical lens through which we can examine the disconnect between what locals and the NPS think about cultural history and the particular management responses to historic events, expanding the existing analysis of these interactions. More specifically, I argue here that metanarratives—by the state, tourists, and to some extent non-Native residents in Wrangell–St. Elias National Park and Preserve—reveal buried epistemologies and create a degree of ongoing violence and displacement, particularly toward Alaska Native peoples.

To be fair, many NPS employees in Wrangell–St. Elias manage cultural resources with good intentions, and local decisions are often made outside their immediate jurisdiction. For example, most of the decisions regarding the large agency investment in Kennecott have been made on regional and national scales. On an individual level, a number of park service employees—in particular the current park subsistence coordinator—are well respected within many park communities and have received awards for their efforts. Nevertheless, as a federal institution and also a regional and local land manager, the NPS does not always realize the politics of narrative and the larger implications of those narratives. Despite individual efforts, in other words, Wrangell–St. Elias National Park is still a national park and is not excepted from the immediate colonial past. In this case, the history plays out in an additional ways by which certain narratives gain prominence and others are erased. The following examination is critical to informing a better understanding of how ongoing colonialist discourse, uneven economic conditions, and race frequently shape contemporary NPS policy.

Background for NPS Management and National Narrative Influence in Alaska

In the late 1970s in Alaska, the National Park Service began the trials of managing what historian Theodore Catton has called "Inhabited Wilderness."[8] The Alaska National Interest Lands Conservation Act of 1980 (ANILCA) set out explicitly to preserve not only the ecological integrity of the land but also the livelihoods and customs of the rural people living on that land. This was an early experiment of co-management in the United States and it applied to both "Native and non-Native people." "It was that tension between the rights of native peoples to be masters of their own cultural evolution on the one hand, and the desires of preservationists to retain the 'primitive' feeling of Alaska's pristine wilderness on the other, which made the new parks so interesting and yet so fraught with difficulty," Catton wrote a decade after ANILCA was passed.

Beginning at the end of the nineteenth century, popular Alaska narratives emerged to encompass the belief that Alaska contained the last pristine wilderness. Describing his Alaska travels, for example, John Muir influenced popular narratives of his time about the role of people in nature: "Civilization and fever, and all the morbidness that has been hooted at me, have not dimmed my glacial eyes, and I care to live only to entice people to look at Nature's loveliness."[9]

What was once fur, gold, and copper was recommodified as "America's Crown Jewel"[10] in the form of wilderness, as many Americans—particularly those living outside Alaska—determined that some of Alaska's vast undeveloped land should be preserved in perpetuity. Ernest Gruening, director of United States territories and later Alaska's governor and a US senator, was the first to recommend that Wrangell–St. Elias be protected. After a flight over the area in 1938, he wrote in a memorandum to Secretary of the Interior Harold Ickes:

> The region is superlative in its scenic beauty and measures up fully and beyond the requirements for its establishment as a National Monument and later as a National Park. It is my personal view that from the standpoint of scenic beauty, it is the finest region in Alaska. I have traveled through

> Switzerland extensively, have flown over the Andes, and am familiar with the Valley of Mexico and with other parts of Alaska. It is my unqualified view that this is the finest scenery that I have ever been privileged to see.[11]

The establishment of Wrangell–St. Elias National Park and Preserve in 1980 confirmed at the national legislative level a new set of representations and meanings that are embodied in the scenic and cultural landscape of the park. The narratives of the 1960s and '70s that envisioned Alaska as part of the last best chance for wilderness became part of the largest wilderness preservation act in the world. ANILCA designated 104 million acres of land in Alaska as wilderness, including the thirteen-million-acre Wrangell–St. Elias National Park and Preserve—an area that exceeds the size of Switzerland and is six times greater than the size of Yellowstone. The act doubled the size of the National Park System with scores of new parks, preserves, and wildlife refuges to the forty-ninth state.

Yet most of the two thousand people who testified in Congress in favor of establishing vast swaths of Alaskan wilderness had never set foot in the state.[12] As Henry Crandall, the chief of the Wilderness Society's legislative program at the time, noted, "The little old lady in tennis shoes in Rhode Island will get the job done in Alaska—not Alaskans."[13] For the more than hundred organizations that were part of the Alaska Coalition at the height of its membership in the 1970s, making Alaska public lands a national issue was a political priority.[14] Today the Alaska Coalition, which was a pivotal advocate for wilderness protection, numbers more than one thousand conservation, sporting, religious, and labor groups. These national visions for Alaska were voiced with moral conviction. Alaska was popularly coined the "last best chance" to protect wilderness and a novel opportunity to extend subsistence lifestyles for both "Native and Non-Native Alaskans."[15]

Despite the promises of ANILCA for the recognition of local livelihoods, the NPS continues to elevate the Kennecott mining story, reinforcing a strong narrative tension between Americans from the lower 48 and park residents. Many aspects of the park that have been constructed by NPS and others to be attractive to tourists are in many cases incommensurate with the history of locals and their interests in the area. Northern frontier narratives surrounding the mining era and the railroad in Wrangell–St. Elias have become smoothed over, "Disneyfied," and normalized in the interest of attracting park visitors. These mining narratives increasingly comply with some of the commercial tourism and development interests of local residents who have benefited from the establishment of the park. In 2013, tourism was the second-largest private-sector employer, accounting for one in eight Alaska jobs.

Tourists, especially out-of-state tourists with the substantial economic means to visit the park, often by air flight, may have preconceived notions about Alaska long before they ever set foot there. These ideas have been drawn from popular literature such as the frontier mining narratives of Jack London and Robert Service, or the early naturalist writings of John Muir, Olaus and Margaret Murie, Robert Marshall, and others.[16] Today, frontier imaginaries may be further fortified by the current onslaught of survival and reality television shows that take place in Alaska, including the recent Discovery Channel reality series *The Edge of Alaska* that was filmed inside the park and incorporated the scripted narratives of many park residents. In other words, NPS narratives are intertwined with other dominant narratives about the North, which have long been imagined by the dominant Western culture as a vast, pristine, sparsely inhabited place. "The last frontier" has always also been imagined as a place of boom and bust, as exemplified by the fur trade in the eighteenth and nineteenth centuries, copper and gold in the late nineteenth and early twentieth centuries, and oil in the present era. Many of these false notions are reinforced rather than expanded or challenged by the NPS.

Absent from the NPS Narrative Frame

While the boom-and-bust Kennecott narrative may be impressive to some, it obscures alternate narratives that describe both the human and the nonhuman world. This narrative only recognizes a particular slice of human history and works conveniently to direct tourists to particular parts of the large protected wilderness area. Just down the hill from the historic Kennecott Recreation Hall, where the centennial celebration took place, the glaciers that parallel the community are rapidly shrinking. During the mining era and well into the 1960s, local residents were not able to see across the panoramic valley and glacial highway that today's tourists marvel at, due to the recent glacial retreat of several hundred feet. Archival photographs too reveal that the physical landscape has changed dramatically. Nevertheless, very little of this visible land change is incorporated in the park interpretive signs in Kennecott. The first sign one views when ascending the hill that precedes the community entrance is a brand-new polished sign marking the Kennecott mining town—in copper.

At the same time, the NPS investment in Kennecott indirectly commemorates the ability of industry, technology, and modernity to overcome "natural" obstacles such as crossing glaciated terrain in the far northern environment, contradicting the agency's mission of preservation, and obfuscating the ongoing living history in the area. Just as wilderness designation has become an ideology based on an imaginary sense of a peopleless landscape, the NPS emphasis on the Kennecott mining history has erased the postmining, pre-park-establishment era and oversimplified a multifaceted ongoing socioecological history in the area. While the height of the industrial age is revered, NPS does not invite scrutiny to the consequences of unmitigated industrialization, and the recent history and human experience of the area remain largely unexamined in NPS literature and interpretation. Based on the current content of the interpretative material offered in Kennecott, one might suppose that the human history of the area largely ceased in 1938, when the Kennecott mines closed, and that little human history preceded the mining era.

The history of mining and human experience in the Wrangells did not end when the last train pulled out of Kennecott in 1938. An account of the period that followed copper mining has yet to be offered in detail by the NPS, however. Understanding the interaction of the area's colonial history and this time period helps us better understand contemporary park identities and relations as well as ongoing disputes over park boundaries and access. NPS does not always recognize the diversity of the social relations within Wrangell–St. Elias Park or acknowledge the intimate relations that both indigenous and nonindigenous people have to each other as well as the nonhuman environment, especially among those who predate the establishment of the park. There is little distinction made between people who have been living in the area for decades or centuries and the recent influx of new arrivals and second homeowners. The period I examine here is one in which white rural Alaskans, often known as "outlaws" as well as "back-to-the-landers," were engaged in similar hunting and subsistence practices as Alaska Natives and often gleaned their skills and knowledge of the land from the Alaska Native communities around them, expanding on the cultural exchanges that occurred during the mining era.

Non-Native Alaska Narratives (1938-2012)

The interim period between the end of mining and the establishment of the national park has been virtually erased by the dominant NPS mining narrative around Kennecott. Following the closure of the copper mines in the late 1930s, the termination of the Kennecott company town, and the quick disintegration of a railroad that was never built to last, virtually all the remaining people who were previously connected to Kennecott became increasingly reliant not only on prospecting and small-scale mining but also on fishing, gardening, hunting, and other subsistence-related activities.

Many of these skills did not develop in isolation from the Alaska Native community. They were strengthened and refined through the interaction of non-Native Alaskans with Alaska Natives in the postmining period. A hunter who has lived in Wrangell–St. Elias since his childhood in the 1960s explains:

> [The Ahtnas] are probably some of the toughest people on the planet. I never encountered anyone as tough as those people . . . they were adapted to this country. They didn't get cold. They were amazing . . . Jack John, he wore lace-up leather boots with a pair of wool socks and usually blue jeans. He would usually have a little wool shirt on and a windbreaker and just your regular monkey face gloves, at 25 to 30 below, you know, and the guy didn't get cold. A lot of them are that way, really woodsman-type people. Jack taught us how to trap and then he taught us how to kill your animals in the trap and . . . how to skin animals, and so you know I owe him a lot, you know what I mean? And you know how to stay dry, how to not sweat, you know there's a lot of things you don't know. They moved

> around real slow and we were like little high-speed people and [Jack John] was just like no. No. Because you're going to sweat up and if you sweat up you are going to die. You know what I mean, because he was used to staying outside all the time and so it was like you moved around out there, you moved slow and you just loaded up to stay warm, but not enough to sweat and then it was interesting. You never got wet, any snow you got on yourself you brushed off immediately, and boy he was on it—you know what I mean—you'd be goofing around in the snow or something, and boy he would just know.[17]

Nevertheless, these stories about the survival lessons that non-Natives learned from Alaska Natives have not resulted in many long-lasting relationships. When Jack John died, members of the Ahtna were angry with this sport hunter for taking far more than he deserved of John's belongings. The hunter's wife had been in charge of the will and said that everything had been left to his son. Films and photographs document this relationship, and the hunter's son shared a number of stories about his experience with Jack John. In addition, Ahtnas have expressed anger that Jack John did not have a Catholic funeral. But according to the sport hunter, he wanted to be cremated and placed next to his father.

There is a historical backdrop to this rift. As copper prices dropped in the 1920s and 1930s, much of the land in what is now considered the national park became a free-for-all for those people who were attracted to notions around a self-sufficient lifestyle, along with those who wanted to live in a space with little government regulation. Failing to fully recognize the lessons and skills they gleaned from Alaska Natives, those who remained in the area, as well as those who streamed in from other parts of Alaska and the lower 48, also soon found supplementary ways to support themselves. They looted windows, cans of food, dishes, and utensils from the former mines. Land was cheap. For example, one old-timer who arrived in the area in 1953 purchased several buildings in downtown McCarthy for less than a hundred dollars.

Another community of primarily Euro-Americans also trickled into the area, not solely for its remaining mineral riches but also for an increasingly rare opportunity: the ghost towns of Kennecott and nearby McCarthy offered the possibility to live a life in a unique natural setting with a departure from contemporary rhythm and pace; many of these newcomers were drawn to the absence of industry. While the railroad aided the import of goods such as oysters, coffee, and pineapple from all over the globe, most of what entered the community in the postmining, prepark era—building materials, food, and drums of gasoline—arrived by air, vehicle, or a small hand-pulled cable car. Quick to make use of the materials that the Kennecott company left behind, these residents devised new uses for the bins that had once transported copper ore from the mine shafts to the crushing mill. In one case, the bins created a novel passage over the Kennicott River following the quick collapse of the railroad bridge, which the Kennecott Company had never built to last. In the words of national parks scholar Joseph Sax, the transformation of the copper bins to a community tram system

> symbolized a degree of self-imposed burden that encouraged people to learn to take care of themselves and to develop their own resources. They were neither hermits or [*sic*] ascetics, but people who wanted to stand aside from the careless ease of pushing a button. One had to develop repair skills or rely on neighbors. And the culture of the tram meant that everyone had to consider pound by pound the cost of hauling the demands of his/her lifestyle.[18]

Though this lifestyle resembles many aspects of Alaska Native culture, there was little recognition of that influence. The above description contrasts sharply with the experience of most community members in the Kennecott area today. Many longtime park residents claim there is less interaction among community members and between people and the harsh features of the nonhuman community in this remote area. While twenty years ago people communicated with CBs over a community radio station, today the phones that were implemented in the mid-1990s have largely been replaced once again by wireless Internet and cell phone reception. In a community that gained its romantic reputation and tourism draw in part for being off the grid, people are now increasingly plugged in. One no longer needs to lug a season's supply of lentils, trail mix, and other nonperishable products from Anchorage or the lower 48. The recent construction of a modern convenience store in the center of town offers imported goods like California-grown avocados and cherries for "Anchorage" prices. Similar to the height of the mining era in Kennecott, once again there is an economy that supports the sale of $4 energy bars, organic cereal, and expensive dried mangoes in the town mercantile. As one old-timer put it, "In the old days we would discuss fixing a wheel on the carts [to transport goods across] the Kennicott River; now we are replacing the carts with no conversation about it."

Many locals of Euro-American descent that I spoke with agree that the interdependence of community members in the past is at risk of being lost to a new form of individualism. Rather than hauling one's construction materials across a raging creek on foot with the help of neighbors or by community-constructed foot bridges that were in constant repair by groups of community members working together, a sturdy private vehicle bridge now sells expensive seasonal vehicle passes for upwards of $500. Loaded in individually owned vehicles, imported bundles of wood from Home Depot regularly cross this bridge to bring construction materials straight to the front door of a cabin. Deliveries of gasoline, imported groceries, and other supplies flow into the community on a daily basis.

Alterations to community living have taken place in less than a decade. Throughout my interviews, non-Native locals who predate the establishment of the park described how they perceive that community relations have changed since the expanded presence of the National Park Service in the late 1990s and the restoration of the Kennecott company mining town. According to one local, "Hardship ensured you could be here. It was the glue that bound people together. People were forced to interact with those they did not like. This has changed in the last ten years. Now it is easier to be here. You do not need to be as nice to your neighbors. We are quickly becoming any park town, USA."[19]

Some residents also express nostalgia for the prepark atmosphere. One female informant told me how, as a child from Fairbanks, she visited the park community of Chitina every summer on family dipnetting excursions on the Copper River. "On the way to Chitina we'd pick up ice to prepare to fish. On the way back we'd resupply with ice from the glacier. Now there is no glacier," she laments. In her early twenties, she went to work on the Alaska pipeline, much of which runs along the boundary of the park.

Though she often worked ten-to-twelve-hour days, seven days a week, she had boundless energy to explore the Copper River Basin, much of which is now national park. "I liked it so much I accepted a layoff in '76 and rented a little cabin, not far from the pump station south of Tonsina. I often refer to it as the best year of my life." She recalls that there were so few people in the park in those days that "you were always glad to run into someone in that big landscape. Running into someone was more of a treat than a problem. Now you go for peace and quiet."

Following the NPS acquisition of Kennecott in 1998, non-Native locals recall that it remained a ghost town for many years. "The park had not done any improvements," she laments. "[There was] mostly original paint, original condition, and nothing visually prominent like there is now. Going there for the [2011] centennial makes me so thankful I was there at the end of the '90s before it blossomed into the different place it is now. Let me retract the word *blossoming*. What does a tumor do?"

Today, many of these residents suggest that the NPS staff has replaced the managers of the company town. And the town has turned into Disney World. They say the agency operates from a distance, most often from Washington, DC, where those making management decisions have never set foot in Alaska, let alone Wrangell–St. Elias Park. They claim most NPS employees leave Kennecott on weekends and return to their more permanent homes in communities outside the park. Millions of dollars have been poured into the ongoing restoration of Kennecott, which has been in process since the early 2000s. Much of that money has stemmed from regional and national NPS offices as well as from private donors. Yet many locals are employed by NPS or private contractors, earning high wages of up to $65 an hour that allow them to remain in the area—long after the summer season. This is the opposite of what went on prior to park establishment, when many local residents left the community in the summer to cash in on lucrative Alaska summer-season industries such as fishing and hunting. Park residents also claim that the NPS investment in Kennecott has created uneven access to the company town or, in today's terms, to the benefits that could result for local residents from park establishment. Kennecott is just one among the twenty-three resident zone communities in the park and other park communities, such as Nabesna and Slana, which are also connected to the road system, that might have benefited from further NPS attention and investment.[20]

One established non-Native hunting guide from the less-traveled Nabesna area spoke to me about the NPS prioritization of Kennecott:

> Yeah, McCarthy, Kennecott—that whole section over there is what they wanted to focus on and that's why they built the ranger station on that side and stuff like that. They had a bunch of drawings they brought out here to show us for this side and they were going to built a road up to the glacier. . . . They were going to build a big park ranger station up there and everything like that and we were kind of against that. It was like you know, really? You really need to do that?[21]

Nevertheless, despite the primary NPS focus on Kennecott, this hunting guide also indicated that his family has been able to maintain their stronghold in the guiding business through park establishment and the increasing regulations that were involved. Today they are the only family with a permit to hunt in the Nabesna area, having squeezed out several other Alaska Native and non-Native guides.

> There has been a huge change from the beginning, you know, the protest marches and all the stuff you know that kind of started way back when, but there's a lot of people now that are actually kind of like me that are looking at it and saying, okay, what are the good points? You know what's happened here and then all of a sudden it's like, okay you know I'm still allowed to be a guide and outfitter. I'm still allowed to be able to hunt, and why it's actually because of the park service. So I have to look at that and so that's changed and I know a lot of people are like, oh they had the park service and whatever, but it's not the "hate the park service," you know what I mean? They have changed, you know what I mean, and a lot more people are saying, well, I'm going to go out and see that park, you know what I mean? And I don't think that—the mountains have always been there, they've always known they've been there, and then they're like, well, it's a park so it must be something special, and I think that's finally starting to set into people and they're starting to say, okay, well it was set aside not

> just to cut me off, you know, and throw me out, it was set aside because it was something special. I think that's starting to click in now to a lot of folks, and you still got your old hard-liners, you know what I mean? You know it was always the deal, "we got to overthrow the park service" and "we got to run them out of this country and no matter what it takes to do it and blah blah blah." You don't hear that much anymore.[22]

Similarly, in another fly-in-only area of the park, a large-game-hunting outfitter expressed a related view of NPS:

> It became a park of course about the same time we were starting to live here, in the '80s [it] became a park and that was kind of a weird time. Both my dad and mom thought you know as far as commercial activities and stuff it was over. Because what he did was hunting mainly but we saw it as recreational opportunities we wanted. I was more into mountain climbing and I guess nonconsumptive tourism, I guess whatever you want to call it, and so we saw a future in that and from day one we basically wanted to get along with the park service. I can't say that any of us were just absolutely thrilled that it was a park but also saw the big picture and realized that it was going to be around a lot longer than we'd even be here so we took the stance initially [of] get along with and invite them over here and say, hey, you know come check us out.[23]

Thus, given a certain degree of privilege, exclusive rights have been gained by some non-Native people who were once opposed to the establishment of the park. With the continued and improved capability to extend their livelihoods following park establishment, these individuals have softened their views and even welcomed the NPS in protecting a valued resource on which their business relies.

Uneven Access to the Company Town: Ahtna Narratives

The National Park Servie could draw on a deeper layer of narratives to remedy the injustices of the past and, particularly, the area's history of colonial displacement. Certain Ahtna narratives offer an important expansion and counterpoint to the above narratives from white landowners. They also offer a different perspective on changes in the land and those caused by NPS, increased tourism, and second homes. Contrary to popular narratives offered by NPS and tourist businesses, prospectors from the lower 48 did not discover copper ore in the Wrangells, and they did not learn the ways of the land on their own.[24] Ahtna people with the longest history in the area and more recent arrivals who are predominantly Euro-American with a short but intense history of living in connection to the land often express different histories about the area. Many residents of Kennecott, for example, deny that there was ever much Ahtna presence in the area, due to the lack of salmon and big game in that particular area of the park. Nevertheless, Ahnta people claim a prominent history in the area, as migratory people and as a culture that relied on copper for its decorative properties. They claim that the richest copper vein in the world was stolen from their chief, challenging the NPS narrative that the copper was spotted on a mountain slope by an Alaska "explorer" after it was initially mistaken for a large patch of grass.

The mining and postmining era allowed for a strong force of resilience among many of the Alaska Native communities in the Wrangell–St. Elias region, who had previously been displaced by colonial influence, physical exile, and cultural indoctrination via forced boarding schools, missionaries, and disease. Many Alaska Natives had provided local provisions to the miners through hunting and gathering, and they shared techniques about how to adapt to a harsh, windy environment where temperatures range from the eighties in the summer to negative forty in the winter. Others were instrumental in the construction of the railroad and filled many jobs that entailed operating heavy machinery. As a result, some Alaska Native families were able to rebound financially and culturally in the era following the mining period, which ended just before World War II. Many entered the sport hunt guiding business that became increasingly popular in remote areas of Alaska in the 1940s and '50s. Local tribal councils remained influential over land rights issues, family relations, and crime. In 1945, Alaska passed some of the country's first civil rights legislation guaranteeing citizens that they would not be discriminated against. Ten years before *Brown vs. the Board of Education*, the Alaska Civil Rights Act outlawed racial segregation and Alaska Natives won equal rights. But in 1968 the discovery of oil in Prudhoe Bay launched the largest and perhaps the most expedited land claims settlement in history.[25]

The 1971 Alaska Native Claims Settlement Act (ANCSA) represented another possible turning point for Native American rights in Alaska that coincided with the civil rights movement. It was unprecedented in the United States for an indigenous population to potentially have so much control over the future of their land. The act divided more than forty million acres of land into twelve geographic regions that would be governed by Native-owned corporations, unlike the reservation system in the lower 48. Yet in section 4, aboriginal title was dissolved and diverse tribal lineages were reduced and forced into federally recognized categories that did not acknowledge their connections to the land of their ancestors—or their experience of colonial history and cultural trauma. Hundreds of individual clans were compressed into state-defined categories. Identity became increasingly measured by blood and corporate accounting.[26] Park preservation, statehood, and native claims were established in just over two decades. These three federal acts were inextricably intertwined, politically, economically, and racially. The process for Native tribes to become federally recognized by ANCSA was pushed through Congress in less than two years.

An interpretation of the hurried, complicated, and disempowering experience of many Alaska Natives is expressed in a series of interpretive letters to the editor that were published in the Native newspaper the *Tundra Times* by Fred Big Jim and James Ito Adler:[27]

> I have been living here in the village for many years . . . we didn't have much to do in the evenings until one day when the mail plane dropped a bundle of magazines, which all turned out to be the same—an ACT. Wally read one copy and told me that it had a lot to do with my future here in Alaska, so we read it together in the evening to practice our [English] lessons. So far it has been pretty one sided because an ACT doesn't have any Eskimo language in it. . . . As there were many new and difficult words in an ACT we read section 3 first on definitions. The very first word defined in an act is "secretary," which does not mean a woman who operates a

typewriter and makes decisions for the boss in an office, as I had always been told. This "Secretary" is the Boss of the interior department. I wonder who writes his letters for him and makes all of his decisions for him like all the other bosses?

Anyway, this secretary seems to be extremely important since he is the first person defined in an ACT and apparently gets to make most of the important decisions. For example, in section 3 of an ACT, it says that a Native is a person that has ¼ native blood, or if not that, then someone who is recognized by other Natives as a Native. Then it says in section 3, any decision of the secretary regarding eligibility for enrollment shall be final. So Wally was wondering if he was a friend of the secretary could he be enrolled as a Native? As for me, I was wondering if the Secretary doesn't like me, could he prevent me from being enrolled as a Native? I mean is a Native a Native or is a Native someone the Secretary says is a Native? How much Native blood does the Secretary need to be able to decide who is Native? And who were the Natives who decided that the Secretary could decide who Natives were.[28]

Prior to the late 1800s, Alaska Native tribes used those areas traditionally controlled by them. Land was generally held by the community as a whole. These boundaries of control were based not on written documents or maps but on traditions and practice. This communal ownership is described technically as "traditional use and occupancy." The political challenge with this type of ownership is that there was no receipt or written title. This was the dilemma that Alaska Natives faced, when, in the late 1950s and early 1960s, the state, corporate interests, conservationists, and sectors of the federal government began encroaching on what was identified by Alaska Natives to be their traditional lands.

A compromise was reached over Alaska Native land claims on December 18, 1971. Native claims in Alaska were unprecedented, yet the access most Alaska Natives received was insufficient in many cases for the continuation of traditional lifestyles, and in most cases was further reduced through park establishment. Today, while half an acre of land in the park near Kennecott sells for over a quarter million dollars, a few of the twenty-three resident zone communities in the park are among the poorest in the nation. Similar to the conditions before ANCSA, the Alaska Native population in this park has the highest rates of alcoholism, infant mortality, and unemployment and lowest cash income and literacy rates in the United States. Scholars have linked such conditions with the expropriation of lands and overturning of traditional communal ownership of land.[29]

Narrative Entanglement

Ten years after the park was established, the horse-packing outfit of one Ahtna family went out of business. As a family friend described, they could not keep up with the ongoing paperwork or competition from other non-Native guides. In addition, following the establishment of the park, the family was no longer able to graze their horses in the area where they had grazed freely in the past. In an interview, one former park employee explained that during the push for park establishment

> Native Alaskans liked the subsistence idea. But they also thought the park would generate Native Alaskan jobs and that there would be wage labor for Natives in the Wrangells. That has never been realized.[30]

One Ahtna man lamented the fact that the Ahnta community was not provided more opportunity to develop tourism in the Nabesna area. "NPS has failed to invest in small microbusinesses that would really benefit Ahtnas," he argues. While this man contends that the National Park Service made a big deal about the Nabesna area being designated as a Superfund site, he notes that cleaning up the area would have cost far less than what NPS has poured into Kennecott. Yet he recalls that back in the 1970s he had a strong sense that park establishment was going to cause uneven benefits. "Those who already had gold received more gold in their pockets," he claims.

The park's current approach (as exemplified in the case of the rail centennial celebration) reinforces a local and national frontier narrative through the boom-and-bust mining story and the symbolic meanings that are attached to that simplified story: mining represented an enormous transition for Alaska Native communities in the park, and though the implications were mixed, in some cases the industry created opportunities for resilience. Today, however, NPS management represents the uneven influence of local residents whose voices were the loudest when Wrangell–St. Elias National Park was established. Early NPS documents refer to much of the (Euro-American) prepark hunting and guiding community as a "fraternity." In a 1981 letter written by the superintendent of Wrangell–St. Elias, newly appointed park rangers were instructed to "avoid lighting a media fire"[31] and to compliment local guides and hunters for a successful season. Local NPS staff was prohibited from engaging in subsistence activities: "We will be under extreme public scrutiny. We cannot appear to have conflict of interest. Our words and actions will be examined critically for real or imagined sensitivity."

From the very inception of Wrangell–St. Elias National Park, local people in rural Alaska communities tormented rangers. In 1978 local protests swept across the region, and effigies of President Carter, who had initially established the area as a preserve through the Antiquities Act, were set to fire. Local resistance to the NPS was strong, from bulldozer operator, hunter, and back-to-the-lander alike, as captured in John McPhee's bestselling book about that era in Alaska, *Coming into the Country*.[32]

These narratives remain powerful and influential today among NPS staff and local community members. Memories of the experience of early NPS employees were evoked during an NPS panel in the summer of 2012, when a number of former NPS rangers described the inhospitable atmosphere of the park. Current and former NPS officials recalled not being allowed to purchase gas or food or find places to live in communities in and around the park. As addressed earlier in this chapter, today many of the locals who were initially uncongenial and unreceptive to the park service now work closely with the agency. Some local residents, primarily non-Native people, have concluded that it is better to work for NPS than against the agency. On the whole, NPS officials are less timid and many of them live in the park year-round with the encouragement of the agency to participate in subsistence activity.

Yet back in the 1970s, some voices were less welcome to the negotiation tables that were involved in park establishment. Though they mostly supported park preservation at that time, today many Ahtna people have little sympathy for these battle stories that NPS officials and locals use to describe the early days of the park. According to an Ahtna man I spoke to, the fierceness of the pre-ANILCA debate for Alaska Natives was never captured by the media and rarely showed up in meeting reports and research materials. He has limited sympathy for the plight of the early NPS staff, as he feels that the hostility he encountered around ANCSA in 1972 was far greater than that experienced by the park employees following ANILCA. "They do not wear the same scars and bullet holes that I do," he said.[33]

He points critically to park managers who take "a one-dimensional snapshot of a four-dimensional" reality. "The one-dimensional snapshot only justifies the stance of those in power." He also contends that the NPS and certain park residents have an ongoing tendency to "take history and use history in terms of forming goals or mission. If the history doesn't fit their mission they disregard it."[34] While he does not cast individual blame, he contrasts the static reductive tendency of park managers with Ahtna understandings. According to the Ahtna people, all things are relational and humans and nature are not separate. They do not perceive any distinct separation between people, the animals, the soil, and the water—or the histories we tell of them. This informant proposes that these false distinctions have been solidified through the English language, which he refers to as "a language of property." Such distinctions may have been further fortified through NPS management policy.

Employment during the mining era had allowed many Ahtna families to resettle the land that had been part of their cultural tradition for thousands of years and adapt to new social and ecological circumstances. In the 1950s, another wave of Americans from the lower 48—referred to by an Ahtna informant as "outlaws"—arrived primarily to engage in small mining operations and to run hunting guide businesses. In that time period several Ahtna families were slowly squeezed out of the guiding business, making displacement more inevitable with park establishment. One Ahtna informant contends that "non-Native outlaw families" stole their land from his family during the 1950s. These "outlaw" narratives gained the most traction with NPS and the pro-park environmental community during Alaska national park establishment:

> When NPS was established in 1980, we had a whole series of park superintendents. First there were the rangers and supers, a few deployed came in and they ended up doing the victimization thing. They considered themselves victims and in the process of considering themselves victims, those who set out to use [the] park for [their] own benefit became their friends . . . the remnants of outlaw society—those people subverted NPS to protect their interests . . . Understand that from where I was coming from these people were outlaws, where you are coming from they are heroes because they did not [violently] oppose NPS.[35]

Ahtna narratives challenge both the NPS and non-Native park resident narratives. The creation of the park, which followed ANCSA, has not benefited the Ahtna at nearly the same scale as many Euro-American families, who have been able to take better advantage of tourism, real estate development, and speculation. While the politics of park establishment rested on the narratives of the environmental movement from the lower 48—that they would protect Native subsistence lifestyles—Ahtna communities were not afforded the same opportunity to take advantage of tourism. "We wanted to start a trekking company, but we could not afford the insurance," an Ahtna told me.[36] An Ahtna informant explained that all the places his mother once used to fish are now privatized. "My mom quit fishing because she has no driver's license. It costs hundreds of dollars to buy a fish rack and haul fish."[38]

While the Ahtna Nabesna narrative takes place in the past, it continues to resonate with contemporary debates about colonial encounters, park history, climate change, and local knowledge. Narratives surrounding local knowledge and Alaska Native displacement have faded in Wrangell–St. Elias due to the increased prominence of current narratives by non-Native Alaskans and state policy. The prioritization of tourism and the rehabilitation of the Kennecott company mining town echo the NPS master narrative that ranks the experience of guests over the area's more comprehensive dynamic living history. While many non-Native Alaskans continue to speak vociferously about being displaced by the establishment of the park, many of them have in fact benefited from the park establishment—and from the ongoing erasure of Alaska Native narratives. This perpetuates a form of symbolic violence where significant histories remain buried. This Ahtna narrative poignantly maintains that colonialism is not something relegated to the distant past in Alaska. The irony is that Alaska Natives are the very people who have shaped, and coevolved with, the so-called wilderness of Alaska. Thus, the contemporary wilderness narrative that the NPS promotes and depends upon buries the role of Alaska Natives in contributing to the production of the environment of Alaska.

Conclusion

How the National Park Service frames the narrative of Wrangell–St. Elias matters not only for park residents but also for a popular understanding of Alaska. The current NPS narrative fits within oversimplified narratives of Alaska. This is especially relevant as a robust popular literature and growing series of reality television further essentializes and romanticizes Alaska as a landscape with exaggerated folk tales about miners, outlaws, and outsiders. Yet these simplified stories contribute little to our understanding of the dynamic human experience in this place, or of specific narratives between Native and non-Native humans and the nonhuman environment, and how these relationships change based on extended time in this place.

In Wrangell–St. Elias National Park, the brief Kennecott mining history has become the baseline narrative around which other park narratives revolve. Concentrating tourism in Kennecott might be a wise strategy for a federal agency that is tasked with managing and preserving the biggest national park in the United States. Nevertheless, this chapter has examined why a history of conquest cannot be easily transferred into an enterprise of historic and ecological preservation. The NPS and local tourism focus on mining extends well beyond the railroad celebration. Kennecott has become a postage stamp for the park, attracting visitors from all over the world. Images of Kennecott often now parallel those of Denali, glaciers, salmon, and grizzly bears in the Anchorage airport. Although the majority of tourists to this 13.2-million-acre park and preserve visit Kennecott based on the advice offered on the NPS website and tourist brochures, the former mining town is just one of the twenty-three diverse resident zone communities in this vast national park. Yet by popularizing Kennecott as a place of free enterprise and industrial development, NPS plays a role in suppressing the vast array of human experience and relationship to the land in the park at large, historically and in the present. NPS also indirectly legitimizes some cultural practices such as tourism over others such as subsistence.

By expanding and complicating the Kennecott narrative, the NPS has an opportunity to better address the impact of more complicated, interactive, and layered narratives within the park. In the Long Range Interpretive Plan[39] for the park, based on legislation and congressional records, special resource studies, and park planning documentation, the "significance" of the park is defined to include the following:

- Wrangell–St. Elias National Park and Preserve is part of one of the largest international terrestrial protected areas in the world, conserving geological and hydrological processes and biological diversity.

- Because of its vast size, high peaks, glaciers, and wilderness, Wrangell–St. Elias National Park and Preserve provides world-class opportunities for discovery, reflection, unconfined recreation and adventure in a remote setting.

- Wrangell–St. Elias National Park and Preserve recognizes that people have been a part of the environment for thousands of years. This park is part of a worldwide conservation trend that recognizes the value of people continuing to live as part of protected places.

- Wrangell–St. Elias National Park and Preserve preserves, manages and presents a collective human heritage which continues to profoundly influence regional, national, and international political and economic developments.

Greater attentiveness to nonmining narratives would allow the agency to better meet the mission of protecting these resources. In addition, as included in this chapter, many local resident narratives note significant ecological changes in the land that can be of great use to NPS management in our biggest national park, which currently employs only one permanent year-round ecologist. Perhaps the American public could learn how locals experience climate change rather than solidifying their image through popular media portrayals of stranded walrus and polar bears due to arctic ice melt. And perhaps this would create more empathy and behavioral change from the public at large.

Yet more deeply missing from the predominant mining narrative in Wrangell–St. Elias Park and Preserve is any perspective on what it is like to have had your culture erased, reestablished, and reimagined as a static representation and then reconsumed as entertainment, economic opportunity, or part of another culture's celebration.[40] Sometimes, as in the case of the railroad centennial, a popular narrative falls apart and demands that we recognize where it falls short, where we have omitted the full story and simplified the plot. Narrative simplification has occurred beyond the conceptualization and framing of human history in Wrangell–St. Elias Park and Preserve. The transformation of this vast area into a national park and, in turn, into a tourist commodity has not only contributed to revised narratives about the area's human history but reinforced a violent colonial history, rather than acknowledging the opportunities for resisting that history. As the political ecologist Alice Kelly argues, once a place "has been attached with new value and economic meaning, it may be only a matter of time before the new fictitious commodity becomes normalized in the imagination."[41]

How we see the history of a place—and people's relation to that place—changes who partakes in the management of that place, how management is understood by the public, and the goals embedded in that management. Accommodating a more inclusive narrative of our largest national park would require abandoning the search for a simple unitary model, particularly the idea of a baseline historical moment such as Kennecott, around which all other histories revolve. The problem with the kinds of simplifications I have discussed in this chapter is that the narratives being simplified are not entirely clear to policy makers and land managers before they too simplify—basing the rationale for one level of abstraction upon the next. A different response to this oversimplification is to pay more attention to the interaction between historic and contemporary narratives in shaping the management of cultural resources. By paying more attention to the interaction of local and national narratives, a new space for imagining alternative narratives might evolve. And this might return more benefits to the Alaska Native community. Perhaps managers of protected areas and cultural histories might better conceive of these models if they tried to build our understanding of how to manage the system based on a greater attentiveness to the interaction of particular narratives at local, national, and global scales. Understanding how people make sense of their interactions with the nonhuman environment and with one another, by paying greater attention to the interaction of park narratives, might allow for more effectual, just, and more widely accepted policy and management interventions to be made.

Acknowledgments

I am grateful for research support from the residents of Wrangell–St. Elias, UC Berkeley fellowships, the George Melendez Wright Society, the Murie Science and Learning Center, the National Park Service, and the U.S. Forest Service. I am also thankful for the archival assistance I received from the Bancroft Library at UC Berkeley, the Denver Public Library, The Wrangell Mountains Center, and the National Park Service in Anchorage and Copper Center, as well as critical feedback that I received from Louise Fortmann's and Nathan Sayre's lab group, particularly from Sibyl Diver, Chryl Corbin, Carolina Prado, Juliet Lu, Maywa Montenegro, Annie Shattuck, Jennifer Durant, Jennifer Bacca, Jeff Martin, and Liz Carlisle.

Endnotes

1. The Kennecott Company name is spelled with an "e" as is the historic town of Kennecott while the Kennicott Glacier and Kennicott River are spelled with an "i." Kennicott is also the frequent spelling used among residents of that community.

2. See D. M. Spence, *Dispossessing the Wilderness: Indian Removal and the Making of the National Parks* (New York: Oxford University Press, 1999), and A. Runte, *National Parks: The American Experience* (New York: Taylor Trade Publishing, 2000), R. Solnit, *Savage Dreams* (Berkeley: University of California Press, 2000), A. Chase, *Playing God in Yellowstone* (New York: Mariner Books, 1987), M. Dowie, *Conservation*

Refugees: The Hundred-Year Conflict Between Global Conservation and Native Peoples (Cambridge, MA: MIT Press, 2009), and others.

3. B. Willems-Braun, "Buried Epistemologies: The Politics of Nature in (Post) Colonial British Columbia," *Annals of the Association of American Geographers* 87, no. 1 (1991): 3–31.

4. J. Kosek, *Understories: The Political Life of Forests in Northern New Mexico* (Durham, NC: Duke University Press Books, 2006).

5. J. Cruikshank, *Life Lived Like Story: Life Stories of Three Native Yukon Elders* (Lincoln: University of Nebraska Press, 1991).

6. C. Finney, *Black Faces, White Spaces: Reimagining the Relationship of African Americans to the Great Outdoors* (Chapel Hill: University of North Carolina Press, 2014).

7. See Spence, *Dispossessing the Wilderness;* Runte, *National Parks;* Solnit, *Savage Dreams*; and Chase, *Playing God in Yellowstone*, among others.

8. T. Catton, *Inhabited Wilderness: Indians, Eskimos, and National Parks in Alaska* (Albuquerque: University of New Mexico Press, 1997).

9. J. Muir, *Travels in Alaska* (Berkeley: Sierra Club, 1915).

10. S. Haycox, *Frigid Embrace: Politics, Economics, and Environment in Alaska* (Corvallis: Oregon State University Press, 2002).

11. E. Gruening, "Memorandum to the Secretary of the Interior," Washington, DC, November 7, 1938, History files, WRST.

12. All interviews were conducted in confidentiality and the names of the interviewees are withheld. Phone interview conducted by Margot Higgins on December 11, 2012.

13. J. Turner, *The Promise of Wilderness* (Seattle: University of Washington Press, 2012), 144.

14. Ibid.

15. I use the term "Native" not only because it is the official term in the federal legislation but also because that is the term most of the indigenous Alaskans I spoke to use to refer to themselves.

16. S. Kollin, *Nature's State: Imagining Alaska as the Last Frontier* (Chapel Hill: University of North Carolina Press, 2001).

17. Respondent #2, interviewed by Margot Higgins at his home in Alaska on July 18, 2012.

18. J. Sax, "Keeping Special Places Special: McCarthy-Kennecott and the Wrangell-St. Elias National Park: A Great Challenge, a Unique Opportunity," Wrangell Mountains Center white paper, 1990.

19. Respondent #3, interviewed by Margot Higgins at his or her office on June 21, 2011.

20. In recent years, the Nabesna area has received some improvements, including the establishment of a new campground on Ahtna land. Nevertheless, this has required a very small portion of the funding that has been allocated to Kennecott.

21. Respondent #4, interviewed by Margot Higgins at his home in Alaska on July 18, 2012.

22. Ibid.

23. Respondent #5, interviewed by Margot Higgins at his home in Alaska on July 2, 2012.

24. See Henry T. Allen, *Report of an Expedition to the Copper, Tananá, and Kóyukuk Rivers: In the Territory of Alaska, in the Year 1885* (Washington, DC: US Government Printing Office, 1887).

25. Haycox, *Frigid Embrace.*

26. J. Smith, "Locating Alaska in Ethnic Studies: Blood, Land and Capitalism," presentation at Perspectives on Native Landscapes Conference, University of California, Berkeley, February 11, 2014.

27. The Alaska Native newspaper *Tundra Times* was founded in 1962 as a means to connect distant and isolated Native communities. It quickly became an influential statewide newspaper read by Natives and non-Natives. For more information on the *Tundra Times*, see Elizabeth James, "Toward Alaska Native Political Organization," *The Western Historical Quarterly* 41, no. 3 (2010): 285–303.

28. F. Big Jim and F. Ito-Adler, *Letters to Howard: An Interpretation of Alaska Land Claims* (Anchorage: Alaska Methodist University, 1974).

29. See Mikyta Daugherty, William James, Craig Love, and William Miller, "Substance Abuse among Displaced and Indigenous Peoples," in *Changing Substance Abuse Through Health and Social Systems*, ed. William Miller and Constance Weiner (New York: Springer, 2002), 225–239, and A. Garcia, *The Pastoral Clinic: Addiction and Dispossession along the Rio Grande* (Berkeley: University of California Press, 2010).

30. Respondent #6, telephone interview by Margot Higgins, October 4, 2014.

31. Wrangell St.–Elias Historic Administrative Records, 1959–1999 ACC WRST-00214, National Park Service Headquarters, Copper Center, Alaska,

32. John McPhee, *Coming Into the Country* (New York: Farrar Straus and Giroux, 1977).

33. Respondent #6, telephone interview by Margot Higgins on February 13, 2013.

34. Ibid.

35. Ibid.

36. Respondent #7, telephone interview by Margot Higgins on March 12, 2014.

37. This data is available in Demographic Profile for Alaska, 2010 U.S. Census.

38. Respondent #8, telephone interview by Margot Higgins on May 17, 2014.

39. National Park Service, *Long Range Interpretive Plan for Wrangell–St. Elias National Park and Preserve*, prepared by the Department of the Interior and Harper's Ferry Interpretive Planning, 2005.

40. For a very relevant essay on the subject, see Ernestine Hayes, "What Will We Do With Our Histories," lecture given at the Egan Library, University of Alaska Southeast, November 2013, and available online at https://www.youtube.com/watch?v=EtsYZzLI2uc (accessed October 4, 2016).

41. A. B. Kelly, "Conservation Practice as Primitive Accumulation," *Journal of Peasant Studies* 38, no. 4 (2011): 683–701.

Bibliography

Allen, H. T. *Report of an Expedition to the Copper, Tananá, and Kóyukuk Rivers: In the Territory of Alaska, in the Year 1885*. Washington, DC: US Government Printing Office, 1887.

Big Jim, F., and F. Ito-Adler. *Letters to Howard: An Interpretation of Alaska Land Claims.* Alaska Methodist University, 1974.

Catton, T. *Inhabited Wilderness: Indians, Eskimos, and National Parks in Alaska.* Albuquerque: University of New Mexico Press, 1997.

Chase, A. *Playing God in Yellowstone.* New York: Mariner Books, 1987.

Cruikshank, J. *Life Lived Like Story: Life Stories of Three Native Yukon Elders.* Lincoln: University of Nebraska Press, 1991.

Daugherty, Mikyta, William James, Craig Love, and William Miller. "Substance Abuse among Displaced and Indigenous Peoples." In *Changing Substance Abuse Through Health and Social Systems*, edited by William Miller and Constance Weiner, 225–239. New York: Springer, 2002.

Dowie, M. *Conservation Refugees: The Hundred-Year Conflict Between Global Conservation and Native Peoples*. Cambridge, MA: MIT Press, 2009.

Finney, C. *Black Faces, White Spaces: Reimagining the Relationship of African Americans to the Great Outdoors.* Chapell Hill: University of North Carolina Press, 2014.

Garcia, A. *The Pastoral Clinic: Addiction and Dispossession along the Rio Grande.* Berkeley: University of California Press, 2010.

Gruening, E. "Memorandum to the Secretary of the Interior." Washington, DC, November 7, 1938, History files, WRST.

Haycox, S. *Frigid Embrace: Politics, Economics, and Environment in Alaska*. Corvallis: Oregon State University, 2002.

Hayes, Ernestine. "What Will We Do With Our Histories." Lecture given at the Egan Library, University of Alaska Southeast, November 2013, and available online at https://www.youtube.com/watch?v=EtsYZzLI2uc (accessed October 4, 2016).

Kelly, A. B. "Conservation Practice as Primitive Accumulation." *Journal of Peasant Studies* 38, no. 4 (2011): 683–701.

Kollin, S. *Nature's State: Imagining Alaska as the Last Frontier*. Chapel Hill: The University of North Carolina Press, 2001.

Kosek, J. *Understories: The Political Life of Forests in Northern New Mexico*. Durham, NC: Duke University Press Books, 2006.

McPhee, J. *Coming into the Country*. New York: Farrar Straus and Giroux, 1977.

Muir, J. *Travels in Alaska*. Berkeley: Sierra Club, 1915.

National Park Service. *Long Range Interpretive Plan for Wrangell–St. Elias National Park and Preserve*. Prepared by the U.S. Department of the Interior and Harper's Ferry Interpretive Planning, 2005.

Runte, A. *National Parks: The American Experience*. New York:Taylor Trade Publishing, 2000.

Sax, J. "Keeping Special Places Special: McCarthy-Kennecott and the Wrangell–St. Elias National Park: A Great Challenge, a Unique Opportunity." Wrangell Mountains Center white paper, 1990.

Smith, J. "Locating Alaska in Ethnic Studies: Blood, Land and Capitalism." Presentation at Perspectives on Native Landscapes Conference, University of California, Berkeley, February 11, 2014.

Solnit, R. *Savage Dreams: A Journey into the Hidden Wars of the American West*. Berkeley: University of California Press, 2000.

Spence, M. D. *Dispossessing the Wilderness: Indian Removal and the Making of the National Parks*. New York: Oxford University Press, 1999.

Turner, J. *The Promise of Wilderness*. Seattle: University of Washington Press, 2012.

Willems-Braun, B. "Buried Epistemologies: The Politics of Nature in (Post) Colonial British Columbia." *Annals of the Association of American Geographers* 87, no. 1 (1991): 3–31.

About the Contributors

Allison K. Athens is an instructor of writing composition and literature at the University of California, Santa Cruz. She completed her PhD in literature with a designated emphasis in feminist studies in 2013. She specializes in northern narrative practices, focusing on stories that explore and transgress the boundaries of language, nation, gender, and species.

Kurtis Boyer is a doctoral candidate in the Department of Political Science at Lund University, Sweden. His doctoral research critically engages with the concepts of human empathy and cognition and the role they play in informing our conceptions of, and relationship to, animals in politics. He has authored a chapter titled "The Limits of Species Advocacy" in the book *Animal Politics and Political Animals* (Palgrave, 2014).

Carly Dokis is an assistant professor in the Department of Sociology and Anthropology at Nipissing University and the author of *Where the Rivers Meet: Development and Participatory Management in the Sahtu Region, Northwest Territories* (University of British Columbia Press, 2015). Her research interests include political ecology, anthropology of development, and collaborative research methodologies. She has worked with communities in northern Ontario and the Northwest Territories with a broad focus on the social, economic, and political consequences of participatory environmental management.

Will Elliott is professor of English at University of Alaska Southeast. His research takes an environmental humanities approach to twentieth- and twenty-first-century

American literature, with a specialization in environments and cultures of the Far North, informed by ecopoetics, discard studies, and posthumanism.

Russell Fielding is a geographer with a focus on human-environment interactions, primarily in coastal or insular settings. He is currently an assistant professor in the environmental studies program at the University of the South in Sewanee, Tennessee. Fielding is interested in questions of subsistence, cultural tradition, and resource conservation. Since 2005 he has been involved in a study of artisanal whaling traditions throughout the Atlantic, with field sites in the Faroe Islands, Newfoundland, and the islands of St. Vincent and St. Lucia in the Caribbean.

Cheryl J. Fish is researching and writing on Sami film, including an essay on Liselotte Wajstedt's film and installation, *The Lost One*. Her essay on Buckminster Fuller and June Jordan's "architextual" collaboration in Harlem appeared in the journal *Discourse*; she has also published essays on environmental justice, activism, and ethnicity in the work of Judith Helfand and Ruth L. Ozeki, and on teaching environmental justice ecocriticism. Her poetry manuscript, *Crater and Tower*, addresses succession in the aftermath of the eruption of the Mount St. Helens volcano and the terrorist attacks of 9/11. She is professor of English at Manhattan Community College, CUNY, and docent in the Department of World Cultures at the University of Helsinki, Finland.

Peter Hemmersam is an associate professor at the Institute of Urbanism and Landscape at the Oslo School of Architecture and Design. He is an architect and his research focus is on urban design, sustainable cities, and new digital technologies in the city. He is part of the Future North project (www.futurenorth.no) that studies northern landscapes under transformation.

Margot Higgins is a visiting assistant professor of environmental studies at Macalester College. She earned her PhD from the Environmental Science, Policy, and Management Department at University of California, Berkeley. Since 2004, she has spent time in Wrangell–St. Elias Park as a scholar, college instructor, and director of the Wrangell Mountains Center, where she serves on the board of directors.

Susan Kollin is a professor of English at Montana State University and was the College of Letters and Science distinguished professor for 2011–2014. Kollin is the author of *Nature's State: Imagining Alaska as the Last Frontier* (University of North Carolina Press, 2001), and her research interests include popular Western film and fiction, transnational American studies, and environmental humanities. She recently edited *A History of Western American Literature* (Cambridge University Press, 2015).

Janike Kampevold Larsen is a literary scholar and an associate professor at the Institute of Urbanism and Landscape at the Oslo School of Architecture and Design. She directs the Tromsø Academy of Landscape and Territorial Studies and is the project leader of the Future North project (www.futurenorth.no).

Kevin Maier is associate professor of English and chair of the Department of Humanities at the University of Alaska Southeast. His essays have appeared in *The Hemingway Review, ISLE: Interdisciplinary Studies in Literature and Environment, Studies in American Naturalism*, and the collection *Hemingway in Context*. He is editor of an anthology on environmental pedagogy titled *Teaching Hemingway in the Natural World* (Kent State University Press, 2017).

John Miller is a lecturer in nineteenth-century literature at the University of Sheffield, having previously worked at the Universities of Glasgow, Edinburgh, East Anglia, and Northern British Columbia. His first monograph, *Empire and the Animal Body* (Anthem, 2012), explores the representation of exotic animals in Victorian and Edwardian adventure fiction. His second book is the coauthored volume *Walrus* (with Louise Miller), published in the Reaktion Animal series (2014). He is the editor (with Susan McHugh and Bob McKay) of the book series Palgrave Studies in Animals and Literature and is currently working on *Fur: A Literary History*.

Daniel Monteith is associate professor of anthropology at the University of Alaska Southeast, where he has been teaching since 1999. He has lived and conducted ethnographic and ethnohistorical research in Southeast Alaska for almost twenty-five years. His work and interests include tribal histories of some of the southern Southeast Tlingit tribes, including the Sanya Kwaan (Cape Fox tribe) and Taanta Kwaan (Tongass tribe). Recently he has been working with the community of Saxman to help them have their subsistence status reinstated by the Federal Subsistence Board.

Sarah Jaquette Ray is associate professor of environmental studies and program leader of the environmental studies program at Humboldt State University. She is author of *The Ecological Other: Environmental Exclusion in American Culture* (University of Arizona Press, 2013) and coeditor, with Jay Sibara, of *Disability Studies and the Environmental Humanities: Toward an Eco-Crip Theory* (University of Nebraska Press, 2017). She is currently working on a book project on affect, action, and inclusion in environmental studies curricula.

Chie Sakakibara is an assistant professor of environmental studies at Oberlin College. As a cultural geographer, her teaching and research interests lie in the field of the human dimensions of global environmental change among indigenous and marginalized communities, specifically on their cultural resilience and socio-environmental justice. She has been collaborating with the Iñupiaq community of arctic Alaska since 2004 to study climate change and cultural survival, and is an adopted member of the two whaling crews in Barrow, Alaska.

Elspeth Tulloch is an associate professor of Canadian literature at Université Laval in Québec City, Canada. Recent work has appeared in the *European Journal of English Studies* and in collections such as *Double-Takes: Intersections between Canadian Literature and Film, Place and Replace: Essays on Western Canada*, and *West of Eden: New Approaches in Canadian Prairie Literature*. She guest edited a special issue on Québec cinema and other arts for the journal *Nouvelles vues sur le cinéma québécois: Revue sur les pratiques et les théories du cinéma du Québec.*

Kyndra Turner researches literature, film, and ecocriticism, and earned her PhD from the Department of English at Arizona State University in 2015. Her dissertation was titled "From *Frankenstein* to *District 9*: Ecocritical Readings of Classic and Contemporary Fiction and Film."

Index

Page numbers with an n denote an endnote; *f* refers to a figure. Bolded page numbers refer to a page range for an entire chapter.